总主编 伍江　副总主编 雷星晖

王欢文　王雪峰　著

# 新型纳米结构材料的设计合成及其电容性能研究

Design and Synthesis of Novel Nanostructured Materials for Supercapacitors

## 内 容 提 要

本书采用脉冲激光沉积、水热、化学气相沉积和电化学沉积等方法设计合成了一系列具有新型纳米结构的超级电容器电极材料,同时对材料的结构、形貌、形成机理进行了研究。实验所得不同结构材料有：多孔氧化镍薄膜、二氧化锰纳米片阵列、氧化镍/石墨烯泡沫、多孔氮掺杂碳纳米管、钴酸镍纳米线和纳米片/碳布、卷曲的石墨烯纳米片、石墨烯/二氧化钒纳米带复合物凝胶、三氧化二铁/石墨烯复合凝胶、石墨烯凝胶、二氧化钛纳米带阵列。

### 图书在版编目(CIP)数据

新型纳米结构材料的设计合成及其电容性能研究 / 王欢文，王雪峰著. —上海：同济大学出版社，2018.10

（同济博士论丛／伍江总主编）
ISBN 978 - 7 - 5608 - 6963 - 6

Ⅰ.①新… Ⅱ.①王… ②王… Ⅲ.①纳米材料－结构材料－研究 Ⅳ.①TB383

中国版本图书馆 CIP 数据核字(2017)第 093347 号

### 新型纳米结构材料的设计合成及其电容性能研究
王欢文　王雪峰　著

| | | | | |
|---|---|---|---|---|
| 出 品 人 | 华春荣 | 责任编辑 | 蔡梦茜　卢元姗 | |
| 责任校对 | 谢卫奋 | 封面设计 | 陈益平 | |

| | |
|---|---|
| 出版发行 | 同济大学出版社　　www.tongjipress.com.cn |
| | （地址：上海市四平路1239号　邮编：200092　电话：021-65985622） |
| 经　　销 | 全国各地新华书店 |
| 排版制作 | 南京展望文化发展有限公司 |
| 印　　刷 | 浙江广育爱多印务有限公司 |
| 开　　本 | 787 mm×1092 mm　1/16 |
| 印　　张 | 11.5 |
| 字　　数 | 230 000 |
| 版　　次 | 2018年10月第1版　2018年10月第1次印刷 |
| 书　　号 | ISBN 978 - 7 - 5608 - 6963 - 6 |
| 定　　价 | 56.00元 |

本书若有印装质量问题,请向本社发行部调换　　版权所有　侵权必究

# "同济博士论丛"编写领导小组

组　　　长：杨贤金　钟志华

副　组　长：伍　江　江　波

成　　　员：方守恩　蔡达峰　马锦明　姜富明　吴志强
　　　　　　徐建平　吕培明　顾祥林　雷星晖

办公室成员：李　兰　华春荣　段存广　姚建中

# "同济博士论丛"编辑委员会

**总 主 编：** 伍 江

**副总主编：** 雷星晖

**编委会委员：**（按姓氏笔画顺序排列）

| | | | | | |
|---|---|---|---|---|---|
| 丁晓强 | 万 钢 | 马卫民 | 马在田 | 马秋武 | 马建新 |
| 王 磊 | 王占山 | 王华忠 | 王国建 | 王洪伟 | 王雪峰 |
| 尤建新 | 甘礼华 | 左曙光 | 石来德 | 卢永毅 | 田 阳 |
| 白云霞 | 冯 俊 | 吕西林 | 朱合华 | 朱经浩 | 任 杰 |
| 任 浩 | 刘 春 | 刘玉擎 | 刘滨谊 | 闫 冰 | 关佶红 |
| 江景波 | 孙立军 | 孙继涛 | 严国泰 | 严海东 | 苏 强 |
| 李 杰 | 李 斌 | 李风亭 | 李光耀 | 李宏强 | 李国正 |
| 李国强 | 李前裕 | 李振宇 | 李爱平 | 李理光 | 李新贵 |
| 李德华 | 杨 敏 | 杨东援 | 杨守业 | 杨晓光 | 肖汝诚 |
| 吴广明 | 吴长福 | 吴庆生 | 吴志强 | 吴承照 | 何品晶 |
| 何敏娟 | 何清华 | 汪世龙 | 汪光焘 | 沈明荣 | 宋小冬 |
| 张 旭 | 张亚雷 | 张庆贺 | 陈 鸿 | 陈小鸿 | 陈义汉 |
| 陈飞翔 | 陈以一 | 陈世鸣 | 陈艾荣 | 陈伟忠 | 陈志华 |
| 邵嘉裕 | 苗夺谦 | 林建平 | 周 苏 | 周 琪 | 郑军华 |
| 郑时龄 | 赵 民 | 赵由才 | 荆志成 | 钟再敏 | 施 骞 |
| 施卫星 | 施建刚 | 施惠生 | 祝 建 | 姚 熹 | 姚连璧 |

袁万城　莫天伟　夏四清　顾　明　顾祥林　钱梦騄
徐　政　徐　鉴　徐立鸿　徐亚伟　凌建明　高乃云
郭忠印　唐子来　闫耀保　黄一如　黄宏伟　黄茂松
戚正武　彭正龙　葛耀君　董德存　蒋昌俊　韩传峰
童小华　曾国荪　楼梦麟　路秉杰　蔡永洁　蔡克峰
薛　雷　霍佳震

秘书组成员：谢永生　赵泽毓　熊磊丽　胡晗欣　卢元姗　蒋卓文

# 总 序

在同济大学110周年华诞之际,喜闻"同济博士论丛"将正式出版发行,倍感欣慰。记得在100周年校庆时,我曾以《百年同济,大学对社会的承诺》为题作了演讲,如今看到付梓的"同济博士论丛",我想这就是大学对社会承诺的一种体现。这110部学术著作不仅包含了同济大学近10年100多位优秀博士研究生的学术科研成果,也展现了同济大学围绕国家战略开展学科建设、发展自我特色,向建设世界一流大学的目标迈出的坚实步伐。

坐落于东海之滨的同济大学,历经110年历史风云,承古续今、汇聚东西,秉持"与祖国同行、以科教济世"的理念,发扬自强不息、追求卓越的精神,在复兴中华的征程中同舟共济、砥砺前行,谱写了一幅幅辉煌壮美的篇章。创校至今,同济大学培养了数十万工作在祖国各条战线上的人才,包括人们常提到的贝时璋、李国豪、裘法祖、吴孟超等一批著名教授。正是这些专家学者培养了一代又一代的博士研究生,薪火相传,将同济大学的科学研究和学科建设一步步推向高峰。

大学有其社会责任,她的社会责任就是融入国家的创新体系之中,成为国家创新战略的实践者。党的十八大以来,以习近平同志为核心的党中央高度重视科技创新,对实施创新驱动发展战略作出一系列重大决策部署。党的十八届五中全会把创新发展作为五大发展理念之首,强调创新是引领发展的第一动力,要求充分发挥科技创新在全面创新中的引领作用。要把创新驱动发展作为国家的优先战略,以科技创新为核心带动全面创新,以体制机制改革激发创新活力,以高效率的创新体系支撑高水平的创新型国家建设。作为人才培养和科技创新的重要平台,大学是国家创新体系的重要组成部分。同济大学理当围绕国家战略目标的实现,作出更大的贡献。

大学的根本任务是培养人才,同济大学走出了一条特色鲜明的道路。无论是本科教育、研究生教育,还是这些年摸索总结出的导师制、人才培养特区,"卓越人才培养"的做法取得了很好的成绩。聚焦创新驱动转型发展战略,同济大学推进科研管理体系改革和重大科研基地平台建设。以贯穿人才培养全过程的一流创新创业教育助力创新驱动发展战略,实现创新创业教育的全覆盖,培养具有一流创新力、组织力和行动力的卓越人才。"同济博士论丛"的出版不仅是对同济大学人才培养成果的集中展示,更将进一步推动同济大学围绕国家战略开展学科建设、发展自我特色、明确大学定位、培养创新人才。

面对新形势、新任务、新挑战,我们必须增强忧患意识,扎根中国大地,朝着建设世界一流大学的目标,深化改革,勠力前行!

万 钢

2017年5月

# 论丛前言

承古续今,汇聚东西,百年同济秉持"与祖国同行、以科教济世"的理念,注重人才培养、科学研究、社会服务、文化传承创新和国际合作交流,自强不息,追求卓越。特别是近20年来,同济大学坚持把论文写在祖国的大地上,各学科都培养了一大批博士优秀人才,发表了数以千计的学术研究论文。这些论文不但反映了同济大学培养人才能力和学术研究的水平,而且也促进了学科的发展和国家的建设。多年来,我一直希望能有机会将我们同济大学的优秀博士论文集中整理,分类出版,让更多的读者获得分享。值此同济大学110周年校庆之际,在学校的支持下,"同济博士论丛"得以顺利出版。

"同济博士论丛"的出版组织工作启动于2016年9月,计划在同济大学110周年校庆之际出版110部同济大学的优秀博士论文。我们在数千篇博士论文中,聚焦于2005—2016年十多年间的优秀博士学位论文430余篇,经各院系征询,导师和博士积极响应并同意,遴选出近170篇,涵盖了同济的大部分学科:土木工程、城乡规划学(含建筑、风景园林)、海洋科学、交通运输工程、车辆工程、环境科学与工程、数学、材料工程、测绘科学与工程、机械工程、计算机科学与技术、医学、工程管理、哲学等。作为"同济博士论丛"出版工程的开端,在校庆之际首批集中出版110余部,其余也将陆续出版。

博士学位论文是反映博士研究生培养质量的重要方面。同济大学一直将立德树人作为根本任务,把培养高素质人才摆在首位,认真探索全面提高博士研究生质量的有效途径和机制。因此,"同济博士论丛"的出版集中展示同济大学博士研究生培养与科研成果，体现对同济大学学术文化的传承。

"同济博士论丛"作为重要的科研文献资源，系统、全面、具体地反映了同济大学各学科专业前沿领域的科研成果和发展状况。它的出版是扩大传播同济科研成果和学术影响力的重要途径。博士论文的研究对象中不少是"国家自然科学基金"等科研基金资助的项目，具有明确的创新性和学术性，具有极高的学术价值，对我国的经济、文化、社会发展具有一定的理论和实践指导意义。

"同济博士论丛"的出版，将会调动同济广大科研人员的积极性，促进多学科学术交流、加速人才的发掘和人才的成长，有助于提高同济在国内外的竞争力，为实现同济大学扎根中国大地，建设世界一流大学的目标愿景做好基础性工作。

虽然同济已经发展成为一所特色鲜明、具有国际影响力的综合性、研究型大学，但与世界一流大学之间仍然存在着一定差距。"同济博士论丛"所反映的学术水平需要不断提高，同时在很短的时间内编辑出版110余部著作，必然存在一些不足之处，恳请广大学者，特别是有关专家提出批评，为提高同济人才培养质量和同济的学科建设提供宝贵意见。

最后感谢研究生院、出版社以及各院系的协作与支持。希望"同济博士论丛"能持续出版，并借助新媒体以电子书、知识库等多种方式呈现，以期成为展现同济学术成果、服务社会的一个可持续的出版品牌。为继续扎根中国大地，培育卓越英才，建设世界一流大学服务。

伍 江

2017年5月

# 前　言

当今世界对化石燃料的过度依赖不但使这一有限资源即将枯竭,而且也给地球环境带来巨大危害。于是,科学工作者对各种清洁或绿色能源技术寄予了前所未有的厚望。超级电容器,是近年来出现的一种新型储能器件,它以其超高比功率和良好循环寿命而著称,但较低的能量密度一直是制约其应用的瓶颈。到目前为止,该问题的解决途径主要包括:① 构建特定取向的纳米结构,使其具有比本体材料更优异的储能性质,例如一维(纳米线,纳米管,纳米带,纳米纤维等)、二维(纳米片,纳米壁等)、三维多层次孔状结构;② 设计复合材料,体现各种组分的协同性能,例如碳/金属氧化物、碳/导电聚合物、碳/氧化物/聚合物。本书在广泛文献调研的基础上,利用脉冲激光沉积等方法设计合成了一系列新型纳米结构电极材料,并研究了其电容性能,主要内容概括如下:

(1) 在氧气气氛下,脉冲激光溅射镍金属靶,产生的等离子体羽流与氧分子反应生成氧化镍,直接沉积在导电基底上。通过采用 X 射线衍射(XRD)、拉曼光谱(Raman spectra)、场发射扫描电镜(FESEM)对复合物进行物理表征,所形成的氧化镍薄膜呈现多孔结构,有利于电解质离子传输。电化学测试表明,该氧化镍薄膜电极具有较好的电容特性，在 1 A·g$^{-1}$ 时，比电容达 835 F·g$^{-1}$。在电流密度高达 40 A·g$^{-1}$ 时，电容保持率为 59%（相对于 1 A·g$^{-1}$）。此外，该氧化镍薄膜电极具有良好的循环寿命，在 1 000 次充放电循环后，电容仅衰减了 6%。

(2) 在臭氧气氛下，脉冲激光溅射锰金属靶，产生的等离子体羽流与臭氧分子反应生成二氧化锰，直接沉积在导电基底上。材料表征结果显示，二氧化锰薄膜以超薄(10 nm)纳米片阵列形式垂直于基底。该二氧化锰纳米片阵列可以直接作为超级电容器电极，电化学测试表明，该阵列电极具有优异的倍率性能（在 100 A·g$^{-1}$ 时，电容保持率为 52%）；较高的比电容（在 1 A·g$^{-1}$ 时，比电容为 337 F·g$^{-1}$）；良好的循环寿命（在 6 000 次循环后电容没有衰减）。

(3) 利用脉冲激光沉积技术，将氧化镍薄膜直接沉积在三维高导电性的石墨烯泡沫上，由于石墨烯的高导电性和泡沫的多孔性非常有利于快速的电子传导和离子传输，该氧化镍/石墨烯泡沫电极在三电极体系中具有很高的电容值(2 A·g$^{-1}$ 时为 1 225 F·g$^{-1}$)和出色的倍率性能（在 100 A·g$^{-1}$ 时，电容保持率为 68%）。在 KOH 水溶液中，以氧化镍/石墨烯泡沫作为正极，多孔氮掺杂的碳纳米管为负极，构建了一种新型的不对称超级电容器。在 0.0~1.4 V 的电位窗口下，当功率密度为 700 W·kg$^{-1}$ 时，能量密度高达 32 Wh·kg$^{-1}$。特别是在 2.8 s 充放电倍率(42 kW·kg$^{-1}$)下，能量密度仍保持有 17 Wh·kg$^{-1}$。同时，该不对称电容器具有优秀的循环稳定性（2 000 次循环后，电容保持率达 94%）。

(4) 利用水热法在柔性的碳布基底上控制合成了钴酸镍纳米线和纳米片。在相同负载量的情况下，纳米线和纳米片呈现出了截然不同的电容性能。测试结果表明，相比于纳米片形貌，纳米线形貌具有更高的比电容值和循环性能。这种不同形貌之间的对比揭示了电化学能量储存中的"过程—结构—性质"的关系。

(5) 利用 Hummers 法将多壁碳纳米管同时沿着横向和纵向剪切为卷曲的石墨烯纳米片。这种卷曲的石墨烯片具有一维纳米管和二维石墨烯的杂化结构。电化学测试表明，相比于原始的多壁碳纳米管，卷曲的石墨烯片在酸性、碱性和中性电解质中均表现出更高的电容性能。例如，在 $0.3 \text{ A} \cdot \text{g}^{-1}$ 时，比电容在碱性电解质中达到 $256 \text{ F} \cdot \text{g}^{-1}$。剪切后电容的增加，主要归因于较高的电解质润湿性、缺陷密度和比表面积。同时，第 6 章也提供了一种大规模合成石墨烯的途径。这种卷曲的石墨烯片有望应用于其他领域，例如传感器，电池材料和气体储存等。

(6) 以五氧化二钒和氧化石墨为原料，通过一步法合成了三维石墨烯/二氧化钒纳米带复合物凝胶。在凝胶形成过程中，一维二氧化钒纳米带和二维石墨烯片通过氢键自组装成交联多孔的微结构。由于这种多孔凝胶结构和纳米带赝电容贡献，石墨烯/二氧化钒纳米带复合物凝胶在 $-0.6\sim0.6 \text{ V}$ 电位窗口下，比电容在 $1 \text{ A} \cdot \text{g}^{-1}$ 时达到 $426 \text{ F} \cdot \text{g}^{-1}$，远大于相同测试条件下的单组分的电容值（$191 \text{ F} \cdot \text{g}^{-1}$ 和 $243 \text{ F} \cdot \text{g}^{-1}$）。此外，由于组分间的正协同效应，复合物凝胶电极表现出更高的倍率性能和循环稳定性。

(7) 为了增加超级电容器的能量密度，现有的研究主要集中于正极材料，而负极材料很少有人研究。在第 8 章的研究中，单晶的三氧化二铁纳米粒子直接生长在石墨烯凝胶上，作为高性能的超级电容器负极材料。在三氧化二铁/石墨烯复合物凝胶形成过程中，三氧化二铁纳米粒子与石墨烯片通过氢键作用自组装形成高表面积多孔结构。在 $-1.05\sim-0.3 \text{ V}$ 电位窗口下，复合物凝胶具有高比电容（在 $2 \text{ A} \cdot \text{g}^{-1}$ 时为 $908 \text{ F} \cdot \text{g}^{-1}$）和优异的倍率性能（在 $100 \text{ A} \cdot \text{g}^{-1}$ 时，电容保持率为 68%）。另外，相比于单纯的三氧化二铁，复合物凝胶的稳定性也有明显的提高。

(8) 为了进一步提高超级电容器能量密度,第 9 章以石墨烯凝胶为正极、二氧化钛纳米带为负极、$LiPF_6/EC-DMC$ 为有机电解质构建了一种新型的杂化电容器。由于石墨烯凝胶的多孔性、高导电性和二氧化钛独特的纳米带阵列结构,杂化电容器有利于离子和电子的快速传输。在 0.0～3.8 V 电位窗口下,能量密度高达 82 Wh·$kg^{-1}$。甚至在 8.4 s 充放电倍率下,能量密度仍保持有 21 Wh·$kg^{-1}$。这些测试结果表明,该杂化电容器具有比超级电容器更高的能量密度和比锂离子电池更高的功率密度。

# 目　录

总序

论丛前言

前言

第1章　绪论 ·············································································· 1

1.1　超级电容器的组成和结构 ················································· 2

1.2　超级电容器的基本原理和分类 ··········································· 3

1.3　超级电容器电极材料的研究进展 ········································ 4

　　1.3.1　碳基材料 ······························································· 5

　　1.3.2　金属氧化物材料 ······················································ 9

　　1.3.3　导电聚合物材料 ···················································· 12

1.4　脉冲激光沉积技术在超级电容器中的应用 ························· 13

　　1.4.1　脉冲激光沉积基本原理 ·········································· 14

　　1.4.2　脉冲激光沉积实验装置 ·········································· 16

　　1.4.3　脉冲激光沉积技术特点 ·········································· 16

1.5　超级电容器性能的测试方法 ············································ 17

　　1.5.1　循环伏安法 ·························································· 17

  1.5.2 恒电流充放电测试 ·················································· 22
  1.5.3 电化学交流阻抗测试 ·············································· 23
  1.5.4 其他测试方法 ······················································ 24
 1.6 纳米尺度下的超级电容器与二次电池的异同点 ················· 25
 1.7 选题依据 ·········································································· 27
 1.8 主要内容 ·········································································· 28

## 第2章 室温下脉冲激光沉积多孔氧化镍薄膜及其高倍率赝电容性质的研究 ·········································································· 30
 2.1 引言 ················································································ 30
 2.2 实验部分 ·········································································· 31
  2.2.1 实验原料与仪器 ··················································· 31
  2.2.2 材料制备 ····························································· 32
  2.2.3 材料表征 ····························································· 32
 2.3 结果与讨论 ······································································ 33
 2.4 本章小结 ·········································································· 36

## 第3章 脉冲激光沉积大面积二氧化锰纳米片阵列及其在超级电容器中的应用 ·········································································· 37
 3.1 引言 ················································································ 37
 3.2 实验部分 ·········································································· 38
  3.2.1 实验原料与仪器 ··················································· 38
  3.2.2 材料制备 ····························································· 38
  3.2.3 材料表征 ····························································· 39
 3.3 结果与讨论 ······································································ 39
 3.4 本章小结 ·········································································· 44

# 目 录

**第 4 章 构建基于氧化镍/石墨烯泡沫和多孔氮掺杂碳纳米管的不对称超级电容器及其超高的倍率性能** ········· 45

4.1 引言 ········· 45

4.2 实验部分 ········· 46

 4.2.1 实验原料与仪器 ········· 46

 4.2.2 材料的制备 ········· 47

 4.2.3 材料表征 ········· 48

4.3 结果与讨论 ········· 49

 4.3.1 正极材料 ········· 49

 4.3.2 负极材料 ········· 54

 4.3.3 不对称超级电容器 ········· 58

4.4 本章小结 ········· 60

**第 5 章 控制生长钴酸镍纳米线和纳米片在碳布上及其不同的赝电容行为** ········· 62

5.1 引言 ········· 62

5.2 实验部分 ········· 63

 5.2.1 实验原料与仪器 ········· 63

 5.2.2 材料的制备 ········· 64

 5.2.3 材料表征 ········· 64

5.3 结果与讨论 ········· 65

 5.3.1 $NiCo_2O_4$ 纳米线和纳米片的合成过程 ········· 65

 5.3.2 材料表征 ········· 66

 5.3.3 电化学表征 ········· 70

5.4 本章小结 ········· 71

第6章 剪切多壁碳纳米管为弯曲石墨烯纳米片及其增强的电容
性能 ······ 72
  6.1 引言 ······ 72
  6.2 实验部分 ······ 73
    6.2.1 实验原料与仪器 ······ 73
    6.2.2 材料的制备 ······ 73
    6.2.3 材料表征 ······ 74
  6.3 结果与讨论 ······ 74
  6.4 本章小结 ······ 81

第7章 三维的石墨烯/二氧化钒纳米带复合物凝胶的制备与电化学
表征 ······ 83
  7.1 引言 ······ 83
  7.2 实验部分 ······ 85
    7.2.1 实验原料与仪器 ······ 85
    7.2.2 材料的制备 ······ 85
    7.2.3 材料表征 ······ 85
  7.3 结果与讨论 ······ 86
  7.4 本章小结 ······ 98

第8章 单晶的三氧化二铁纳米粒子生长在石墨烯凝胶作为超级
电容器负极材料 ······ 99
  8.1 引言 ······ 99
  8.2 实验部分 ······ 100
    8.2.1 实验原料与仪器 ······ 100
    8.2.2 材料的制备 ······ 102

|     | 8.2.3 材料表征 ········································ | 102 |
| --- | --- | --- |
| 8.3 | 结果与讨论 ············································ | 103 |
| 8.4 | 本章小结 ·············································· | 111 |

**第9章 基于石墨烯凝胶正极和二氧化钛纳米带阵列负极的杂化超级电容器的构建及其超高的能量密度** ············ 113

| 9.1 | 引言 ·················································· | 113 |
| --- | --- | --- |
| 9.2 | 实验部分 ·············································· | 114 |
|     | 9.2.1 实验原料与仪器 ································ | 114 |
|     | 9.2.2 材料的制备 ···································· | 114 |
|     | 9.2.3 材料表征 ······································ | 115 |
| 9.3 | 结果与讨论 ············································ | 115 |
| 9.4 | 本章小结 ·············································· | 122 |

**第10章 结论与展望** ············································ 123

| 10.1 | 结论 ················································· | 123 |
| --- | --- | --- |
| 10.2 | 展望 ················································· | 123 |

**参考文献** ······················································ 125

**后记** ·························································· 162

# 第1章
# 绪 论

目前,世界各国都投入了极大的财力、物力和人力发展新型电化学能量转换及贮存技术,并研制出许多高性能的化学电源。其中,超级电容器(Supercapacitors)[1,2],是近年来出现的一种新型储能器件,也是绿色能源技术的重要组成部分。它与目前广泛使用的各种储能器件相比,电荷存储能力远高于物理电容器,充放电速度和效率又优于一次或二次电池(图1-1)。此外,超级电容器还具有对环境无污染、循环寿命长、使用温度范

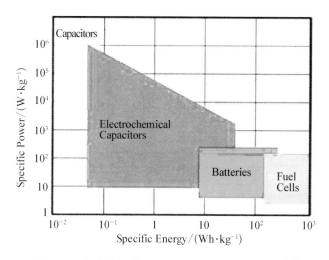

图1-1 各种电化学能量储存装置的能量-功率图[4]

围宽、安全性能高等特点[3]。它与氢动力汽车、混合动力汽车和电动汽车的发展密切相关。采用超级电容器和电池组合的方法,构成混合电源系统作为电动汽车的动力电源可以满足各种技术要求,启动加速时,主要是电容器放电;减速刹车时,可以由制动的充电系统给电容器充电,回收能量。这样可以降低电池的负荷峰值,延长电池寿命,提高能量的利用效率。因此,采用超大容量电容器/电池混合驱动系统被认为是解决电动汽车驱动的最佳方案之一。实际上,超级电容器的用途不仅限于此,它在通信、无线电电子技术、计算机电源、军事、航天领域以及人体医学等方面也有用武之地。

## 1.1 超级电容器的组成和结构

超级电容器和其他电化学储能设备在配置方面存在许多相同之处。通常,超级电容器系统包含正极、负极、电解液、隔膜四个部分(图1-2)。在充电过程中,电解液中的阳离子(例如,$Li^+$,$K^+$,$H^+$等)或阴离子(例如,$OH^-$等)以扩散、迁移、吸/脱附等方式从一个电极运动至其对电极。同时,

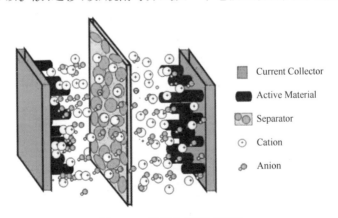

图1-2 超级电容器示意图

电子将沿着外电路从阴极流向阳极。在放电过程中,阴/阳离子从对电极脱出或解吸附,然后通过电解液回到原电极处;电子则从阳极流回阴极,通过外电路并对外做功。工作电极的制造技术、电解质的组成和隔离膜质量对超级电容器的性能有决定性的影响,电解质的分解电压决定超级电容器的工作电压。水溶液电解液电容器的工作电压在 1 V 左右,而有机电解液电容器的工作电压在 3 V 左右[5]。

## 1.2　超级电容器的基本原理和分类

基于电荷存储机理(图 1-3),超级电容器可以分为两类[6,7]:即双电层电容器(Electric double-layer capacitor)和氧化还原电容器(Redox capacitor),后者也称赝(或准)电容器(Pesuocapacitor)。对于前者而言,能量的存储主要来源于高表面电极材料和电解质溶液接触界面处电荷(电子电荷或离子电荷)的有序分离;对于后者,在电极表面或体相中的电活性物

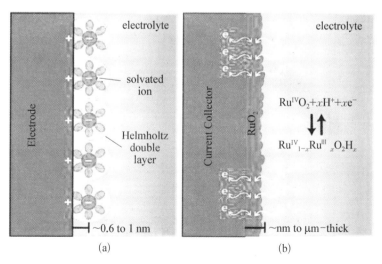

图 1-3　超级电容器电荷存储机理"示意图":(a) 双电层电容　(b) 准电容[4]

质进行欠电位沉积,发生高度可逆的化学吸脱附或氧化还原反应(Faradaic reactions),产生与电极充电电位有关的赝(或准)电容,其储存电荷的过程不仅包括电极活性物质由于氧化还原反应而将电荷储存于电极中,而且包括电极材料表面与电解质之间的双电层电荷的储存。这种电极系统的电压随电荷转移的量呈线性变化,表现出电容的特征,故称为"准电容"。充电时,电解液中的离子(一般为 $H^+$ 或 $OH^-$)在外加电场的作用下由溶液中扩散到电极/溶液界面,而后通过界面的电化学反应而进入到电极表面活性氧化物的体相中;若电极材料是具有较大比表面积的氧化物,这样就会有相当多的电化学反应发生,大量的电荷就被存储在电极中。放电时这些进入氧化物中的离子又会重新返回到电解液中,同时所存储的电荷通过外电路而释放出来[8,9]。

## 1.3　超级电容器电极材料的研究进展

电极材料是超级电容器技术发展的核心(图 1-4),按照电位窗口的不同,电极材料可以分为正极材料和负极材料(图 1-5),用作超级电容器电

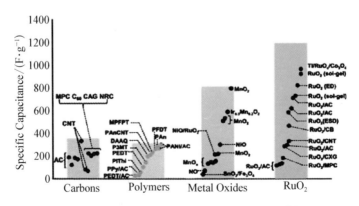

图 1-4　各种超级电容器电极材料的电容值[4]

极材料的物质应具有适当的热力学稳定性,以及良好的电子、离子导电性。目前已经得到商业应用或受到广泛关注的电极材料主要包括三大类:碳基材料、金属氧化物材料和导电聚合物材料。

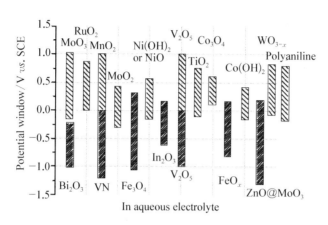

图1-5 一些赝电容材料的电位窗口[10]

## 1.3.1 碳基材料

碳(炭)材料比表面超高,价格低廉,有利于实现工业化大规模生产,早在1957年Becker就提出用活性炭作为双电层电容器的电极材料并申请专利,经过50多年的发展,各种碳基材料(包括活性炭、活性炭纤维、碳气凝胶、碳纳米管以及石墨烯)的制备以及应用技术逐渐趋于成熟。围绕碳材料的比表面积、孔隙结构、表面活性和电子导电性,科学家和工程师开展了许多富有成效的研究工作,其重点主要包括:

(1) 扩充储电空间——高能量密度

碳基电容器的储电机理是电荷在电极表面的有序富集。对于超级电容器而言,有效"比表面积"越大(电解质溶液可以接触和渗透的表面),其储存电荷容量越大。不含缺陷的$sp^2$碳材料的极限比表面积是$2630 \text{ m}^2 \cdot \text{g}^{-1}$(单层石墨烯);而富含缺陷的$sp^2$碳材料的极限比表面积还要大于这个数值。

通常很难获得单层石墨烯片,提高碳材料比表面积的主要方法是营造孔隙,提高表面碳原子的比例,从而增加其比表面积;可是,孔隙率的增加制约了其功率特性的进一步提高。如何在提高比表面积,获得高比电容的同时,保持高的功率密度是获得高性能超级电容器的重要依据。

例如,Ruoff 研究组利用强碱活化的方法处理氧化石墨制备了比表面积高达 3 100 $m^2 \cdot g^{-1}$ 的多孔碳材料(图 1-6),在有机电解质(BMIM $BF_4$)/AN 中,能量密度和功率密度分别达到 70 Wh·$kg^{-1}$ 和 75 kW·$kg^{-1}$[11]。中国科学院金属研究所和澳大利亚昆士兰大学合作提出了可能同时具有高能量密度和高功率密度电容特性的层次孔的思想[12],发展了一种层次孔碳材料的新合成方法,并制备出具有大孔—介孔—微孔三维层次结构和局域石墨片层结构的多孔碳材料(图 1-7)。该材料在大电流密度

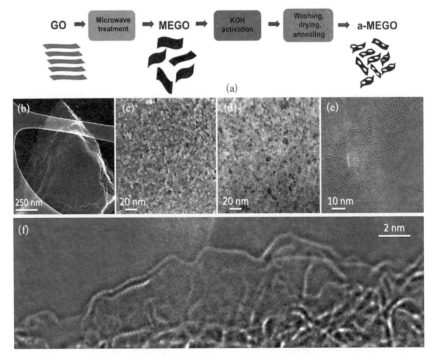

图 1-6 石墨烯材料的处理方法和形貌[11]

条件下具有高能量密度和高功率密度,表明其有望成为超级电容器的优良电极材料。哈尔滨工程大学的范壮军课题组以多孔氧化镁为模板通过气相沉积法生长了均匀的一层石墨烯,当模板在盐酸中溶解后得到了介孔的石墨烯纳米片(图1-8),其比表面积高达1 654 $m^2 \cdot g^{-1}$,在6 M KOH电解质中,扫描速率为500 $mV \cdot s^{-1}$,比电容仍高达202 $F \cdot g^{-1}$[13]。

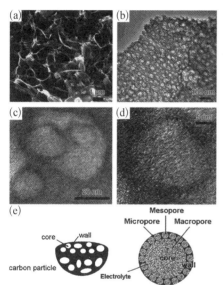

图1-7 三维层次结构的多孔碳[12]

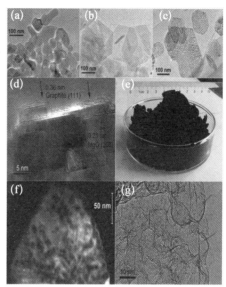

图1-8 三维结构的介孔石墨烯[13]

(2) 控制微观结构和宏观结构——高功率密度

一般来说,通过提高孔隙率可以获得高比表面积碳材料。但孔隙的存在也带来另一个问题,即电解质溶液的有效扩散问题等。如何在提高比表面积的同时,保持其电解质溶液对电荷储存表面的浸润,保证电解质离子以较快的速率从溶液体相向碳材料表面扩散,是碳材料方面需要解决的重要问题之一。例如,美国Drexel大学Gogotsi教授和法国国家科研中心Portet教授合作发现了超微孔中的反常电容现象[14],为多孔碳材料的孔隙结构设计提供了新的思想(图1-9)。

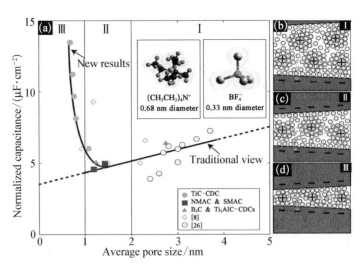

图1-9 孔结构与比电容的关系[14]

(3) 提高石墨烯片层的结构完整性——高导电特性和低内阻

电极材料需要优异的导电特性,完整无缺陷的石墨烯片层具有良好的导电特性。作为电极材料的$sp^2$碳材料应该具有良好的结构完整性。通过活化等方法产生孔隙和缺陷,在提高碳材料比表面的同时,导电性能变差。如何在提高比表面积的同时,不降低$sp^2$碳材料的导电性能也是提高碳电极材料电容性能需要克服的难点[15]。作为$sp^2$碳质材料基元结构的单层(或薄层)石墨烯,是可以突破以上瓶颈的理想材料。其主要原因如下:单层(或薄层)石墨烯片,具有无孔隙的二维平面结构。储电空间位于石墨烯片表面,其储能特性依赖于石墨烯的表面化学和比表面积。微米级的石墨烯片可以自组装形成三维石墨烯宏观体,具有简单的结构特性,不含孔隙,与电解质溶液有良好的接触。经过与其他材料的复合,可以调控其结构,保证材料良好的功率特性。石墨烯片零缺陷或者较少缺陷,可以保证其具有良好的导电和导热特性,可作电极材料,特别是微型的电源所用电极材料的最佳选择。例如,中山大学的沈培康教授课题组通过高温热处理吸附了镍离子的离子交换树脂(图1-10),得到了一种三维的石墨烯网络[16],它的比表面积高达1 810 $m^2 \cdot g^{-1}$,孔容达

1.22 cm³·g⁻¹,其中,介孔和微孔孔容分别为 0.47 cm³·g⁻¹ 和 0.38 cm³·g⁻¹;在 6 M KOH 溶液中,比电容在电流密度为 0.5 A·g⁻¹ 和 100 A·g⁻¹ 时分别为 305 F·g⁻¹ 和 228 F·g⁻¹;在 TEMABF₄/PC 电解质中,比电容在 1 A·g⁻¹ 和 32 A·g⁻¹ 时分别为 178 F·g⁻¹ 和 156 F·g⁻¹。

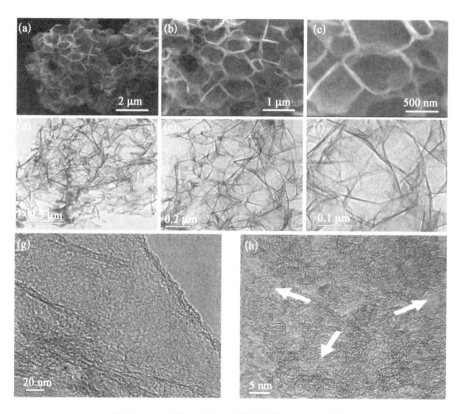

图 1-10 三维多孔石墨烯网络的电镜图片[16]

## 1.3.2 金属氧化物材料

金属氧化物在电极/电解液界面处产生的法拉第赝(或准)电容显著大于碳材料产生的双电层电容,所以对金属氧化物的研究引起了许多研究工作者的兴趣[17,18],最具代表意义的是 RuO₂、IrO₂ 等贵金属氧化物,尤其是 RuO₂ 具有很高的比电容和良好的导电性以及超长循环稳定性,围绕 RuO₂

开展的工作很多。概括而言,研究工作主要集中在以下 3 个方面:① 使用各种方法制备大比表面积的 $RuO_2$ 做电极活性物质;② 把 $RuO_2$ 与其他金属化合物混合,以达到减少 $RuO_2$ 用量的同时又提高电极材料的比容量的目的;③ 寻找其他的廉价材料代替 $RuO_2$ 以降低材料成本。例如,我国台湾国立中正大学的 C. C. Hu 课题组制备了一种用作超级电容器材料的 $RuO_2$ 纳米管[19](图 1-11),该材料的比电容达到了 1 300 F·$g^{-1}$,然而因为钌元素的高昂成本限制了其商业化应用。

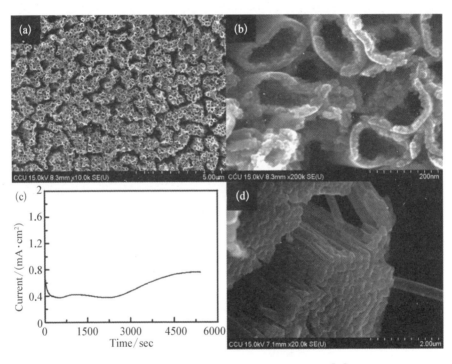

图 1-11　水合 $RuO_2$ 纳米管的扫描电镜图片[19]

为了充分降低电极材料成本,一些科技工作者试图寻找廉价的氧化物电极活性材料来代替 $RuO_2$。例如,日本东北大学的 M. Chen 课题组把二氧化锰沉积在多孔金基底上(图 1-12),二氧化锰的比电容高达~1 145 F·$g^{-1}$,非常接近于理论值[20]。佐治亚理工大学的 M. Liu 组通过水热法把四氧化三钴纳米线沉积在碳纤维纸上(图1-13)[21],在两电极体系中(0.8 V电位窗口下),比电容在25.34 A·g$^{-1}$电流密度下达到1 124 F·g$^{-1}$。Y. H. Lee课题组研究碳纳米纤维纸上电沉积五氧化二钒薄膜(图1-14)[22],当薄膜厚度为3 nm时,氧化钒的比电容值达1 308 F·g$^{-1}$。此外,多孔氧化镍纳米粒子薄膜直接沉积在镍箔上(图1-15)[23],其比电容高达1 487 F·g$^{-1}$。

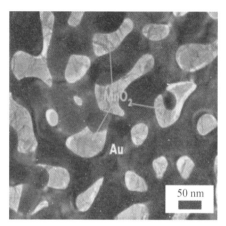

图1-12 多孔金/二氧化锰电镜图[20]

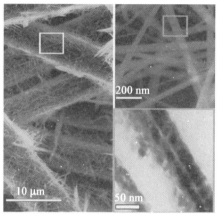

图1-13 四氧化三钴纳米线/碳纤维纸电镜图[21]

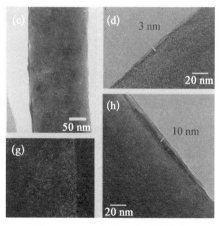

图1-14 五氧化二钒/碳纳米纤维纸电镜图[22]

图1-15 多孔氧化镍纳米粒子薄膜电镜图[23]

### 1.3.3 导电聚合物材料

导电聚合物材料具有良好的电子导电性、内阻小、比容量大等优点,用导电聚合物作为超极电容器电极材料也是近年来发展起来的一个新的研究领域[24]。常用的电极材料有聚吡咯(PPY)、聚苯胺(PAN)、聚并苯

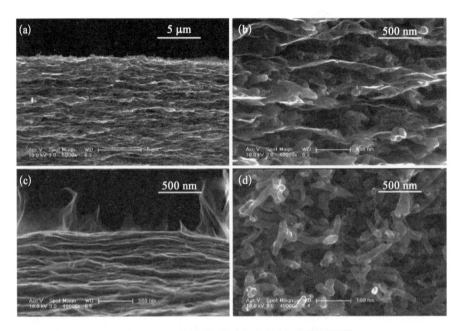

图 1-16 石墨烯/聚苯胺复合材料形貌图片[28]

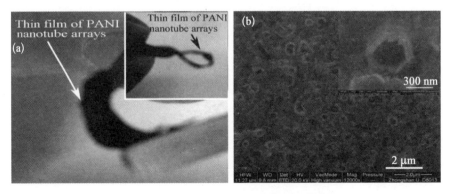

图 1-17 聚苯胺纳米管的形貌图片[29]

(PAS)、聚噻吩(PTH)、聚对苯(PPP)、聚乙烯二茂铁(PVF)等[25-27]。用导电聚合物作为电极材料的电化学电容器的电容主要来自法拉第赝(或准)电容,其作用机理是:通过在电极上的聚合物膜发生快速、可逆的 n 型或 p 型掺杂和去掺杂氧化还原反应,使聚合物达到很高的储存电荷密度,从而产生法拉第赝(或准)电容来储存能量,其较高的工作电位是源于聚合物的导带与价带之间有较宽的能隙。目前仅有限的导电聚合物可以在较高的还原电位下稳定地进行电化学 n 型掺杂,如聚乙炔、聚吡咯、聚苯胺、聚噻吩等,但掺杂时仍存在电阻过大或循环性能不好的问题。现阶段的研究工作主要集中在寻找具有优良的掺杂性能的导电聚合物并进行结构优化,以期提高聚合物电极的充放电性能、循环寿命和热稳定性等。例如,清华大学 G. Shi 课题组制备了基于石墨稀/聚苯胺纳米纤维柔性薄膜材料的超级电容器(图 1-16)[28],电容输出为 210 F·g$^{-1}$。另外,中山大学童叶翔课题组利用模板法构建了一种聚苯胺纳米管[29],比电容在 5 mV·s$^{-1}$ 时为 210 F·g$^{-1}$(图 1-17)。

## 1.4 脉冲激光沉积技术在超级电容器中的应用

根据近年来的文献报道,薄膜材料在超级电容器电极材料中具有十分重要的作用,因为薄膜结构对于电子的传输和离子的渗透具有独特的优势。第一,在传统的电极制备过程中,为了使活性材料紧密固定在集流体上,通常需要加入聚四氟乙烯作为黏结剂。可是,聚四氟乙烯黏结液本身是绝缘的,增加了电极的电阻,不利于电子的传输;另一方面,聚四氟乙烯的存在形成了一定的空间位阻,不利于电解质离子的渗透。然而,在薄膜结构中,由于活性材料直接沉积在集流体上,避免了黏结剂的使用,在充放电过程中,减少了"死体积",增加了电极的导电性。第二,薄膜电极的厚度可以通过控制沉积参数来调节,可以降低电极的等效串联电阻(Equivalent series resistance),从而有利于进行大电流充放电。特别是,在薄膜结构中,形成高度有序的多孔结构,对于离子的渗透非常有效。到目前为止,薄膜材料的制备方法主要有水热法、模板辅助的电沉积法、化学气相沉积法、化学浴沉积法、原子层沉积法、磁控溅射沉积、粒子束溅射沉积、真空蒸发沉积、等分子束外延以及脉冲激光沉积(Pulsed laser deposition,PLD)技术等。其中,脉冲激光沉积作为一种真空物理沉积方法,过程最为简单,沉积速率高且使用范围广。脉冲激光沉积技术始于20世纪60年代初,Smith等人在1965年利用PLD技术制备了多种半导体薄膜、光学薄膜、电介质薄膜等。随着激光技术的发展,PLD被用于制备低温超导薄膜等。在20世纪80年代初,红宝石激光沉积技术可以在高真空室中沉积外延生长的半导体薄膜。随后,PLD技术又被用于制备各种氧化物、氮化物,甚至包括有机-无机复合材料薄膜等领域,在制备金刚石薄膜、磁性薄膜、非线性光学薄膜、立方氮化碳薄膜等领域中也取得了很大进展,同时,PLD在制备纳米颗粒和半导体量子点等领域也有使用。

### 1.4.1 脉冲激光沉积基本原理

当一束高能量的脉冲激光聚焦后照射到靶上时,靶材会被瞬间加热、熔化、汽化,直至变成等离子体,然后等离子体在气氛中从靶表面向基片传输,形成等离子体羽流(Plume),最后定向局域膨胀飞向基底,在基底表面凝聚、成核、长大形成薄膜。脉冲激光沉积过程可大致分为三个阶段,图1-18为此过程示意图:

(1) 激光与物质相互作用产生等离子体

高强度脉冲激光溅射靶材→聚焦处温度急速升高,直至靶材汽化温度以上→靶材熔蚀,汽化蒸发→汽化物质与光波继续互相作用,电离并形成

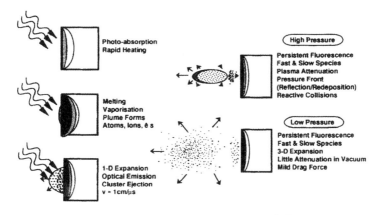

图1-18 脉冲激光沉积过程示意图[29]

区域化的高浓度等离子体→等离子体以新的机制吸收光能,飞向基底。其中,激光汽化固体靶材的过程,可用如下公式来描述

$$\Delta d = (1-R)\tau(I-I_{th})/\rho\Delta H$$

式中 $\Delta d$——汽化深度;

$R$——靶材表面反射系数;

$\tau$——脉冲宽度;

$I$——激光强度;

$I_{th}$——汽化的阈值光强;

$\rho$——靶材的体相密度;

$\Delta H$——靶材汽化潜热。

(2) 等离子体向基片扩散

等离子体火焰在空间的扩散是指激光脉冲结束后,烧蚀物从靶表面到基片的过程。在PLD制备薄膜时经常会有一定压强的气氛气体存在,因此烧蚀物在扩散过程中将通过碰撞、激发、散射以及气相化学反应等一系列过程,而这些过程又影响和决定了在烧蚀物到达基片时的状态、计量比、动能等。

(3) 等离子体中粒子在基片上生长薄膜

上一过程中形成的高能粒子飞向基片表面后迅速冷却→在基片表面键合形成原子团或团簇→随着连续蒸发,这些原子团不断吸收新的原子→生长成为临界核→临界核逐渐生长形成岛状颗粒→岛不断长大,最终形成连续的薄膜。

### 1.4.2 脉冲激光沉积实验装置

脉冲激光沉积系统如图 1-19 所示,包括激光系统、真空沉积膜系统和检测系统三部分。激光系统主要包括激光器和透镜等;真空沉积系统主要包括真空腔、分子泵、机械泵、靶和基片等部分;检测系统主要包括真空计、热电偶、加热控制器等,用来定量控制各种实验参数。

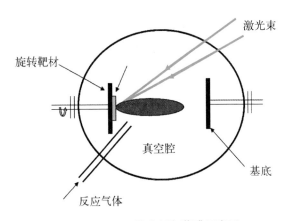

**图 1-19 PLD 技术制备薄膜示意图**

### 1.4.3 脉冲激光沉积技术特点

脉冲激光沉积技术主要优势有:① 设备简单、易于控制,沉积效率很高,可以制备和靶材成分一致的化合物薄膜,甚至含有易挥发元素的多元化合物薄膜,需要调节控制的参数较少而且相对独立;② 激光具有极高的能量,可以瞬间蒸发金属、半导体和陶瓷等无机材料,从而有效地解决难熔材料(如硼化物、硅化物、碳化物、氧化物等)的沉积问题;③ 通过PLD技术可以实现超晶格薄膜、多层复合膜的生长;④ 适用于制备一系列压电、铁电、光电、高温超导等多种功能薄膜,因为PLD可在不同温度下原位生长具有一致取向的织构膜和外延单晶膜;⑤ 在PLD技术中,超高的粒子动能可显著增强二维生长和抑制三维生长,所以利用此方法能够获得超薄的连续薄膜,而且不易出现岛化,即可沉积高质量纳米薄膜;⑥ 由于薄膜在真空室内进行,便于制备原子级清洁的界面;⑦ 易于掺杂,可以直接通过采用的靶材的元素比例进行掺杂。

脉冲激光沉积技术的主要缺陷有:① 在薄膜中及其表面可能存在微米或亚微米尺度的颗粒物;② 所制备的薄膜面积较小。这些缺点随着对脉冲激光沉积技术的深入研究,通过改变激光能量、基片与靶的角度和距离、气体分压等参数将会逐步得到解决。

## 1.5　超级电容器性能的测试方法

### 1.5.1　循环伏安法

在表征超级电容器电化学性能的测试方法中,循环伏安是非常快速而且有效的方法[3-5,11]。在测试过程中,电势从原始电位 $E_0$ 开始,以一定的速率 $v$ 扫描到电位 $E_1$ 后,再反向扫描到原始电位 $E_0$,然后在 $E_0$ 和 $E_1$ 之间进行循环扫描。其施加电势和时间的关系为

$$E = E_0 - vt$$

式中　$E_0$——原始电位(V);

$v$——扫描速度(mV·s$^{-1}$);

$t$——扫描时间(s)。

电势和时间关系曲线如图 1-20(a)所示。电极电势随时间作对称的三角波变化,记录电流随时间的变化情况,可以得到响应电流与电压的关系曲线。因此,循环伏安法能够直观地考察充放电过程中的电化学行为,例如电极反应的难易程度、析氧(氢)特性、可逆性、充放电效率、电极界面的吸/脱附等性质以及在工作电压范围内是否具有理想的电容行为,是否具有稳定的工作电位等。

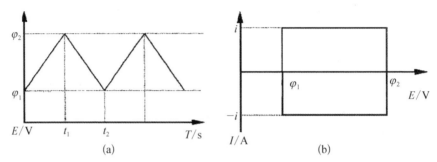

图 1-20　循环伏安测试的给定信号图(a),与响应信号图(b)[11]

采用循环伏安法表征超级电容器时,一般采用三电极体系(工作电极、辅助电极和参比电极),为了模拟超级电容器实际工作的条件,研究内容主要有:

(1) 对于双电层电容器,通常用平板电容器模型进行理想等效处理。平板电容器容量的计算公式为

$$C = \frac{\varepsilon S}{4\pi d} \tag{1-1}$$

式中　$C$——电容(F);

$\varepsilon$——介电常数;

$S$——电极板正对面积,等效双电层有效面积($m^2$);

$d$——电容器两极板之间的距离,等效双电层厚度(m)。

由式(1-1)可知,双电层电容器的容量与双电层的有效面积成正比,与双电层厚度成反比。

（2）对于活性炭电极而言，双电层有效面积与电极的比表面积及电极上碳的负载量有关，双电层厚度则是受到溶液中离子的影响。因此，电极制备好后，电解液确定，电容值基本确定。利用公式 $C=Q/\varphi$ 和 $\mathrm{d}Q=i\mathrm{d}t$ 可得

$$i = \frac{\mathrm{d}Q}{\mathrm{d}t} = C\frac{\mathrm{d}\varphi}{\mathrm{d}t} \tag{1-2}$$

式中　$i$——电流(A)；

$\mathrm{d}Q$——电量的微分(C)；

$\mathrm{d}t$——时间的微分(s)；

$\mathrm{d}\varphi$——电位的微分(V)。

因此，如果在工作电极上施加一个线性变化的电位信号时，得到的电流信号将会是一个不变的量。如果给定的电位信号是一个如图 1-20(a) 所示的三角波信号，电流信号将会是一个恒定的正电流信号或者一个恒定的负电流信号，响应信号如图 1-20(b) 所示，响应信号在 $i$-$\varphi$ 图中呈矩形。

由式(1-2)可知，在扫速不变的情况下，电流($i$)和电容($C$)成正比例关系，即，对于一个电极，在一定扫速下进行循环伏安测试，根据所得曲线纵坐标上电流的变化，就可以通过下面公式计算出电极的电容值：

$$C = \frac{i}{m\left(\frac{\mathrm{d}\varphi}{\mathrm{d}t}\right)} \tag{1-3}$$

式中　$C$——电容(F)；

$i$——电流(A)；

$m$——电极上活性材料的质量(g)；

$\mathrm{d}\varphi/\mathrm{d}t$——扫速($\mathrm{V\cdot s^{-1}}$)。

然而，在实际情况中，超级电容器总会有一定的内阻，它相当于多个静

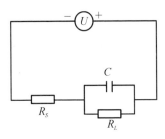

图 1-21 等效电路图

电电容器和电阻混联而成,为了方便,可将其简化为如图 1-21 所示的等效电路,首先由一个电容($C$)与一个等效漏电电阻($R_L$)并联后再与一个等效电阻($R_S$)串联构成。由于 $R_L$ 远大于 $R_S$,又与电容并联,零输入响应(从电压为零开始加载一个定电压 $U$)时可看作断路,此时电路是 $RC$ 一阶回路(以下推导中用 $R$ 代表 $R_S$),可以得到任一时刻的电压为

$$u(t) = i(t)R + u_c \tag{1-4}$$

式中 $u(t)$——等效电路两端的电压(V);

$R$——$R_S$,等效串联电阻(Ω);

$u_c$——电容 $C$ 两端电压(V)。

由

$$i(t) = \frac{dQ}{dt} = \frac{d(C \times u_C)}{dt} = C\frac{du_C}{dt} \tag{1-5}$$

可得

$$\frac{du_C}{dt} = \frac{i(t)}{C} \tag{1-6}$$

因此

$$u_c(t) = \frac{1}{C}\int_0^t i(t)dt \tag{1-7}$$

代入可知

$$u(t) = i(t)R + \frac{1}{C}\int_0^t i(t)dt \tag{1-8}$$

对上式两边分别求导,为

$$\frac{\mathrm{d}u(t)}{\mathrm{d}t} = i'(t)R + \frac{1}{C}i(t)$$

又由
$$v = \frac{\mathrm{d}u(t)}{\mathrm{d}t}$$

可知：
$$i(t) = Cv(1 - \mathrm{e}^{\frac{-t}{RC}}) \tag{1-9}$$

由式(1-9)可知，在电容两端加上线性变化的电压信号时，电路中电流不同于静电电容器那样立刻变化为恒定电流 $i$（循环伏安曲线为矩形），而是需要经过一定的时间。所以，图中循环伏安曲线会出现一段有一定弧度的曲线，式中，$RC$ 为超级电容器的过渡时间。当 $RC$ 较大时，曲线偏离矩形就较大，如图 1-22(a)所示；当 $RC$ 较小时，曲线如图 1-22(b)所示。超级电容器的固有特点是超高的功率密度，这就要求它的内阻很低，从而减小内阻的分压，这时 $RC$ 就很小，曲线非接近于矩形。因此，可通过循环伏安曲线的形状来定性地考察电极材料的电容性能，如果适合作为超级电容器材料，就可利用 $CV$ 曲线上电流的变化来确定电极的比电容值，通过式(1-10)就可以研究比较各种材料的电容性能：

$$C = \frac{i(t)}{mv\,\mathrm{e}^{\frac{-t}{RC}}} \tag{1-10}$$

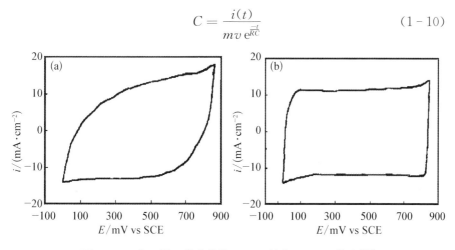

图 1-22 实际循环伏安曲线(a) $RC$ 较大，(b) $RC$ 较小[11]

由式(1-10)可知,在电容值不变的情况下,电流随着扫速的增大而按比例增大;过渡时间却不随扫速的变化而变化,所以当以比电容量为纵坐标时,扫速越快,曲线偏离矩形就越远。如果在较大的扫速下,曲线仍呈现较好的矩形,说明电极的过渡时间小,即电极的内阻小,适合大电流工作;反之,则电极不适合大电流工作,应当采取措施使电极的内阻降低。

### 1.5.2 恒电流充放电测试

恒电流充放电(Galvanostatic charge-discharge)是研究超级电容器电化学性能常用的一种方法[14]。使工作电极在恒电流条件下进行充放电,同时记录电位随时间的变化。本书通过恒电流充放电实验来测试电极材料的电容特性。

可以利用恒流放电曲线通过式(1-11)来计算材料的比电容值[1]:

$$C_m = \frac{it_d}{m\overline{\Delta V}} \qquad (1-11)$$

式中　$t_d$——放电时间(s);

　　　$\overline{\Delta V}$——放电电压降低平均值(V);

　　　$m$——为单电极上活性物质的质量。

实验中采取不同大小的电流密度对电极进行充放电性能测试,研究其功率特性。如果在大电流充放电条件下,电压和时间的曲线仍为线性关系,且计算出的比电容值衰减很小的话,表明具有较好的功率特性。反之,则功率特性较差。

在实际情况中,由于超级电容器存在一定的内阻,由电路学的知识可以得知,充放电转换的瞬间会有一个电位的突变($\Delta\varphi$),如图1-23(b)所示,可以利用这一突变计算电容器的等效串联电阻(ESR)[6]:

$$R = \frac{\Delta\varphi}{2i} \qquad (1-12)$$

式中　$R$——电容器等效串联电阻($\Omega$);

　　　$i$——充放电电流(A);

　　　$\Delta\varphi$——电位突变(V)。

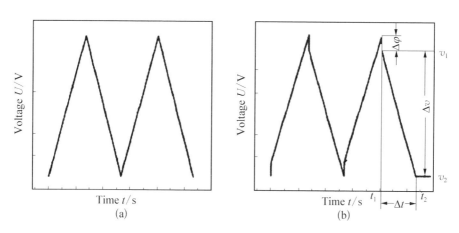

图 1-23　恒电流充放电曲线(a) 理想充放电曲线;(b) 实际充放电曲线

由于有些电极的电容值随着电极电位有一定的变化,因此相应的充放电曲线不完全是直线。可以从电极的恒电流充放电曲线来判断电极的电容性质。理想的充放电性能应该表现为电位随时间线性变化、具有小电流缓慢充电、大电流快速充电的特征。

### 1.5.3　电化学交流阻抗测试

交流阻抗测试(EIS)可以提供研究电极过程动力学及界面结构的重要信息,主要取决于[6]:① 材料的固有性质;② 电极的工艺参数;③ 材料的高比表面积和孔径分布。通过对体系赋予一个小振幅的交流(一般为正弦波)电压(或电流)信号,使工作电极的电位在平衡电位附近进行微扰,从而引起特定的响应信号,由此可以得到有关电极反应动力学的数据。同时,电化学交流阻抗还能测量电极界面/溶液双电层电容和溶液电阻。在测量过程中,交流阻抗不对测试体系产生任何影响,所以特别适用于电极过程动力学机理的研究。

图1-24为EIS测试所得复平面阻抗谱。$Z'$和$-Z''$分别对应于阻抗的实部和虚部。曲线包括高频、中频和低频区域。通常,理想超级电容器不存在频率效应,相应的曲线是一条垂直线。非理想电容器则在高频区域存在一个半圆,对应于电荷传递电阻;中间频率区域是一条呈45°的直线的Warburg区域;在低频区域逐渐过渡为一条直线。在高频区域,因为电解液离子只能进入大的内部空隙,电阻较小;频率降低,电解液离子可以扩散进入电极内部空隙,电容量大幅提高,同时阻抗也迅速增加。在低频区域,如果EIS曲线垂直于$Z'$轴,意味着"电荷饱和",在此频率以下电容器的大部分电容量均可得到利用。此外,可以通过公式(1-11)计算出电容值,

$$Z'' = (2\pi f C)^{-1} \tag{1-11}$$

式中 $Z''$——EIS的虚部值;

$f$——频率;

$C$——电容值。

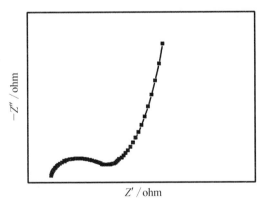

图1-24 一般混合频谱曲线

### 1.5.4 其他测试方法

除了上述三种测试手段,还有漏电流、自放电测试、对超级电容器装置

工艺的评价等。本书主要通过循环伏安、恒电流充放电和交流阻抗测试来考察电极材料的电容性能。

## 1.6 纳米尺度下的超级电容器与二次电池的异同点

超级电容器(特别是赝电容器)与二次电池虽然具有不同的电荷存储机理,但二者均依赖于电化学过程。在过去的5～7年内,由于纳米材料的广泛研究,电化学能量储存领域有了突飞猛进的进展,特别是一些传统意义上的电池材料也被应用于超级电容器[30,31]。此外,一些电池材料在纳米尺度下却表现出电容特性。

一般情况下,对于二次电池(例如锂离子电池),通过锂离子嵌入本体材料而发生的氧化还原反应是一个受扩散控制的过程,其速率很慢;而超级电容器(如双电层电容器)是通过电解质离子在电极表面的吸附来储存电荷的,不涉及氧化还原反应,因而不受制于扩散过程,表现出快速的充放电,具有超高的功率密度,但是,电荷仅限于电极材料的表面,因此,超级电容器的能量密度远低于电池[32]。超级电容的循环伏安图形状接近于矩形,而二次电池的循环伏安曲线具有明显的氧化还原峰,其峰位差通常不低于 0.1～0.2 V。在 20 世纪 70 年代,Conway 等人首次发现了在一些电极材料表面(或内表面)发生可逆的氧化还原反应[33],类似于双电层特征,但具有更高的电荷存储能力,被称为赝电容材料,最典型的例子就是 $RuO_2$ 和 $MnO_2$,近些年又扩展到其他的氧化物、氮化物和碳化物。赝电容器的应用有望改善超级电容器的能量密度,接近于二次电池,但是二者之间的电荷存储机理是完全不一样的。

近年来,纳米材料的广泛应用极易导致赝电容器和二次电池的混淆。

当电池材料处于纳米尺度范围内时,其功率密度也会由于较短的离子/电子传输通道而增加。可是功率密度的增加在大多数情况下并不意味着该类电池材料转变为赝电容器,因为材料的氧化还原峰和充放电曲线仍属于电池类型。但在有些情况下,例如在小于 10 nm 维度内,一些传统的电池材料(如 $LiCoO_2$[34],$V_2O_5$[35])表现出电容性质(图 1-25 上)。这种内在"赝电容"的出现在一些薄膜材料中表现得更为突出。就循环伏安曲线而言,当氧化还原峰的峰位差非常小的时候,即可认为是电容特征。在这种情况下,峰电流密度和扫描速率呈线性关系。最新的文献报道,介孔 $Nb_2O_5$ 也属于类似情况(图 1-25 下)[36,37]。另外,当库伦效率很低且动力学滞后时,也不认为是赝电容行为。

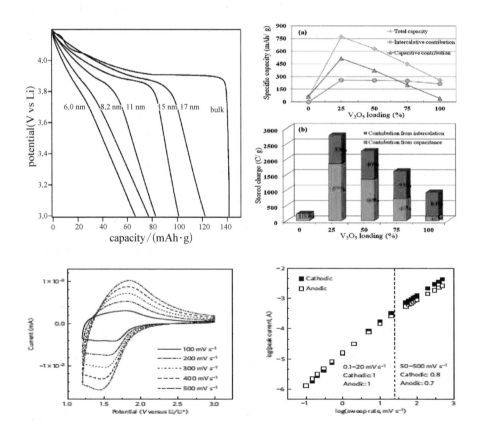

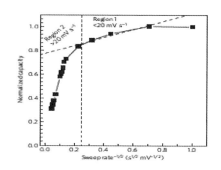
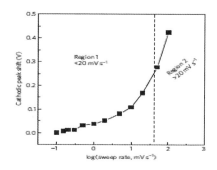

**图 1-25** 电池材料 $LiCoO_2$（左上）[34]，$V_2O_5$（右上）[35]，$Nb_2O_5$（下）[36] 的电化学性质

总之，仅用电池材料在低倍率下的容量值来描述超级电容器性能是没有任何意义的，必须要从高倍率特性、峰位差、峰电流密度与扫描速率的响应关系以及库伦效率和动力学特征来全面界定。

## 1.7 选题依据

在化石能源日渐枯竭、环境污染日益严重、全球气候变暖的今天，寻求替代传统化石能源的可再生绿色能源，谋求人与环境的和谐显得十分重要。于是，科学工作者提出了资源与能源充分利用和环境最小负荷的发展理念，对各种绿色清洁能源储存体系寄予了前所未有的厚望。超级电容器是目前重要的"绿色"储能装置。近年来，纳米材料的广泛研究极大地推动了超级电容器等先进储能技术的蓬勃发展。尽管取得一系列研究进展，但是也存在一些问题，主要有：① 片面追求能量密度而牺牲了超级电容器应有的高功率密度；② 较差的结构稳定性问题；③ 不能同时具有高效的电子和离子传输通道；④ 与电解液接触面有限等。

本书正是针对出现的这些问题，综合考虑高能量密度、高功率密度、轻的质量和优异稳定性等因素（图 1-26），在实验室已有工作和国内外已报

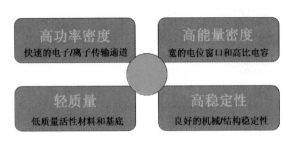

图 1-26 设计高性能超级电容器电极材料的示意图

道最新学术动态的基础上,以提高材料电导率和构建多孔结构,提高比表面为出发点,按照系统科学的实验设计、技术路线和研究思路,设计和合成一系列新型的纳米结构超级电容器电极材料,试图解决相关领域的关键科学问题。同时我们希望书中所述材料可应用于其他领域并对其他材料的结构调控具有指导意义。

## 1.8 主要内容

本书围绕以上选题依据,主要研究内容如下:

(1) 脉冲激光沉积多孔氧化镍薄膜,在碱性电解质中,考察其高倍率赝电容性能;

(2) 利用脉冲激光构建二氧化锰纳米片阵列,在中性电解质中,考察其超级电容性能;

(3) 在 KOH 溶液中,以氧化镍/石墨烯泡沫作为正极,多孔氮掺杂的碳纳米管为负极,构建了一种新型的不对称超级电容器;

(4) 在柔性的碳布基底上控制合成了钴酸镍的两种结构:纳米线和纳米片,揭示了电化学能量储存中的"过程—结构—性质"的关系;

(5) 利用 Hummers 法将多壁碳纳米管同时沿着横向—纵向剪切为卷

曲的石墨烯纳米片,考察其电容性能;

(6) 通过一步法合成了三维石墨烯/二氧化钒纳米带复合物凝胶,考察组分间的正协同效应,进而提高复合物凝胶的倍率性能和循环稳定性;

(7) 单晶的三氧化二铁纳米粒子直接生长在石墨烯凝胶上,作为高性能的超级电容器负极材料;

(8) 以石墨烯凝胶为正极,二氧化钛纳米带为负极,$LiPF_6$/EC-DMC 为有机电解质构建了一种新型的杂化电容器,有效地提高了超级电容器的能量密度。

# 第2章
# 室温下脉冲激光沉积多孔氧化镍薄膜及其高倍率赝电容性质的研究

## 2.1 引　言

在过去的一些年,能量消耗日趋严重,空气污染和全球变暖引起了广泛的关注,并激起了学者对发展高功率密度和高能量密度的能量储存/转化装置的研究兴趣[38,39]。超级电容器作为一种新型的能量储存/转化装置,引起了广泛关注[40]。众所周知,赝电容器由于通过法拉第反应来储存电荷而具有比双电层电容器更高的比电容。赝电容材料主要包括过渡金属氧化物、氮化物和导电聚合物。其中,钌氧化物和锰氧化物是最理想的选项[41],但是,氧化钌的高成本和氧化锰的低导电性限制了其实际应用。氧化镍因其具有高理论比电容、高的化学稳定性、良好的可逆性、环境友好和低成本而受到科研人员的关注[42]。

到目前为止,各种形貌的氧化镍已被应用于超级电容器电极材料,例如多孔微米球[43]、纳米花[44]、纳米片[45]和纳米纤维[46]。不过,和其他金属氧化物一样,氧化镍具有低倍率性能。通常,影响电极材料倍率性能的关键因素有两个[47]:① 电子在电极中的传导;② 离子在电极/电解质界面的

第 2 章　室温下脉冲激光沉积多孔氧化镍薄膜及其高倍率赝电容性质的研究

传输。为了克服和解决这两个关键问题,科研人员通常将电极材料制成薄膜结构,因为该结构不仅可以降低电阻而且能缩短离子扩散的通道。例如 M. S. Wu 组报道了一种具有大孔的氧化镍薄膜,其电容性能得到明显改善[48]。

目前文献报道的制备氧化镍薄膜的技术主要有:电化学沉积[48]、喷雾热分解[49]、热蒸发[50]和化学气相沉积[51]。实际上,脉冲激光沉积(PLD)以原理简单而常应用于薄膜制备和研究[52]。

在本章,主要通过脉冲激光沉积法制备氧化镍薄膜。当该薄膜作为超级电容器电极材料时,表现出高比电容、优异倍率性能和良好稳定性。

## 2.2　实验部分

### 2.2.1　实验原料与仪器

实验过程中用到的原料如表 2-1 所示。

表 2-1　实验原料

| 原料名称 | 生产厂家 | 备注 |
| --- | --- | --- |
| 镍 | Alfa Aesar | 纯度≥99.995% |
| 氧气 | 上海化工院青磊特种气体有限公司 | 纯度≥99.995% |
| 氢氧化钾 | 江苏强盛化学股份有限公司 | 分析纯 |

实验过程中使用的仪器:

(1) 脉冲激光沉积系统:① 真空系统:实验所需低真空由机械泵(TRP-24)产生,高真空由涡轮分子泵(FD-11B)产生,真空室内真空度由电阻规(ZJ-527/KF25)及冷规(ZJ-14/CF35)测量,并以数字形式显示于复合真空机(ZDF-Ⅴ-LED)上。② 光路系统:实验使用的激光器是由 Spectra Physics 公司生产的 Nd：YAG 激光器，为保证激光信号的稳定输出，将该激光器置于隔振平台(RS-1511)之上。为了使激光顺利溅射至金属靶上，采用反射镜(1 064 nm, $R>99.5\%$)，聚焦透镜($f=150$ mm)及相关夹持镜架与连接杆件自行搭建了一套光路用于激光光束的转折与聚焦。③ 配气系统：配气管路由 Swagelok 公司生产的不锈钢管件，三通阀门，精密针阀与自行定制的不锈钢储气罐等组成。

(2) 实验过程中使用的电子天平 BT25S 是由 Sarotius 生产的，电极材料的电化学性质测试使用上海辰华仪器有限公司生产的 CHI660d 电化学工作站。

### 2.2.2 材料制备

在沉积之前，先把体系的压强通过机械泵和分子泵抽至 $10^{-4}$ Pa，随后氧气通过针型阀通入体系中，压强保持在 50 Pa。基底(镀金的钛片)与镍靶之间的距离是 3 cm。将激光聚焦于高纯镍靶材表面上，形成的等离子体羽流与氧气反应，产物沿靶材表面法线方向飞出，碰到距离靶材约 3 cm 处的基底后，在其表面沉积形成氧化镍薄膜。在沉积过程中，靶材通过马达不断旋转，以提高沉积速率和沉积薄膜的均匀性。

### 2.2.3 材料表征

**1. 物理表征**

实验中利用场发射扫描电子显微镜(FESEM；Philips XSEM30, Holland)观测样品表面形貌，通过 X 射线衍射(XRD：D/Max-2400；Cu 靶：$\lambda=1.541\ 8$ Å；管电压 40 kV；管电流 60 mA；扫描速度 5°/min)和显微共聚焦拉曼光谱仪(514 nm laser with RM100)对样品进行结构分析。

#### 2. 电化学表征

上述得到的氧化镍薄膜直接作为工作电极,在室温下采用三电极测试系统在 CHI660d 电化学工作站上进行。所有测试均在 1 M KOH 溶液中进行,以铂片做对电极,饱和甘汞做参比电极。

## 2.3 结果与讨论

脉冲激光沉积的系统如图 2-1 所示。尽管在靶材表面激光融化的过程很复杂,但是其装置是很简单的。当激光束聚焦于旋转的镍靶时,产生的高能粒子(例如,原子簇、离子、分子)与氧分子反应生成氧化镍,沉积在导电基底上。图 2-2(a)为薄膜的 X 射线衍射图,其特征峰完全归属于 NiO(JCPDS, No. 22-1189)。拉曼光谱(图 2-2(b))具有明显的两个峰(约 545 cm$^{-1}$ 和 1 077 cm$^{-1}$),分别属于氧化镍的 1 p 和 2 p 散射[45]。从扫描电镜图片可以看到(图 2-2(c)~图 2-2(e)),氧化镍薄膜呈现疏松的多孔结构,且薄膜的厚度约为 700 nm。这种多孔的薄膜结构具有很高的有效比表面积,有利于电解质离子的传输。

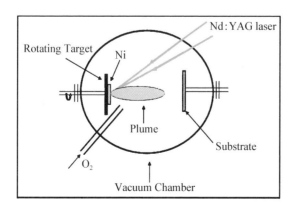

图 2-1 制备氧化镍薄膜的脉冲激光沉积系统

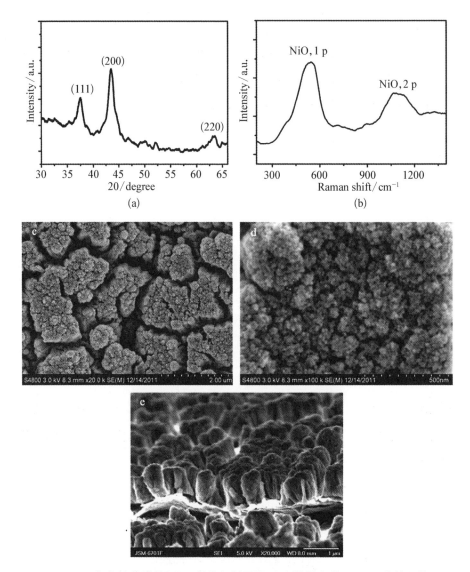

**图2-2** 氧化镍薄膜的(a) X射线衍射图片,(b) 拉曼光谱,(c~e) 电镜图片

多孔氧化镍薄膜电极在碱性电解质中的赝电容行为通过循环伏安和充放电测试得到。图2-3(a)为氧化镍薄膜和基底在 $10\ mV\cdot s^{-1}$ 的循环伏安图。氧化镍薄膜的循环伏安曲线具有明显的一对氧化还原峰,属于赝电容特征,电极反应可表示为:$NiO+OH^- \longleftrightarrow NiOOH+e^-$。另外,也

## 第 2 章 室温下脉冲激光沉积多孔氧化镍薄膜及其高倍率赝电容性质的研究

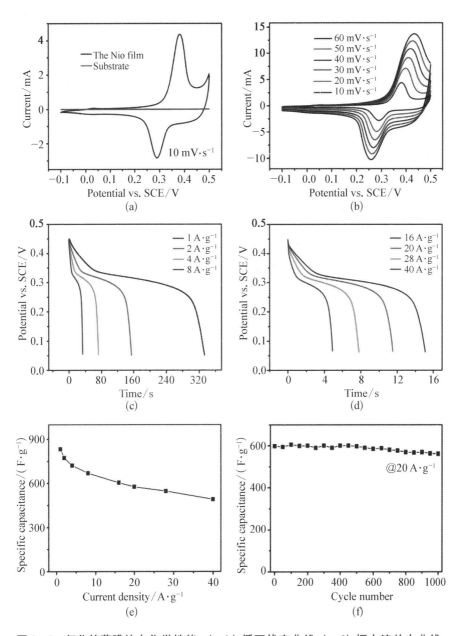

图 2-3 氧化镍薄膜的电化学性能：(a,b) 循环伏安曲线，(c,d) 恒电流放电曲线，(e) 不同电流密度下的电容值，(f) 循环稳定性

可以发现,相比于氧化镍,导电基底的循环伏安曲线几乎是一条直线,由此表明电容的贡献来源于氧化镍。随着扫描速率的增加,电流密度也相应地增加,但是循环伏安曲线的形状基本保持不变,说明氧化镍薄膜具有良好的倍率特性。根据放电曲线,可以计算出不同电流密度下的比电容值。在电流密度为 1 A·$g^{-1}$ 时,比电容达 835 F·$g^{-1}$。在电流密度高达 40 A·$g^{-1}$ 时,电容保持率为 59%(相对于 1 A·$g^{-1}$)。此外,该氧化镍薄膜电极具有良好的循环寿命,在 1 000 次充放电循环后,电容仅衰减了 6%。

理解氧化镍薄膜电极的高比电容和倍率特性的本质是很有价值的。在脉冲激光沉积的过程中,激光产生的粒子具有超高的动能和激发能(达几千电子伏),反应生成的氧化镍与导电基底之间具有极强的结合力,有利于电子从薄膜到基底之间的快速转移;相比于本体材料,准二维的薄膜材料具有较低的电荷转移电阻(ESR);此外,薄膜本身的多孔结构提供了有效的比表面积,从而增加了电解质离子的传输通道。因此,该氧化镍薄膜具有优异的超电容性能。

## 2.4 本章小结

在室温下,利用脉冲激光沉积成功制备了多孔氧化镍薄膜电极,该薄膜的多孔结构十分有利于电荷的储存,进而展现出很高的比电容、倍率性能和优异的稳定性。本章在电化学储能领域提供了一种新的方法——脉冲激光沉积法(PLD)。

# 第3章
# 脉冲激光沉积大面积二氧化锰纳米片阵列及其在超级电容器中的应用

## 3.1 引 言

构建具有高比表面积和有效离子/电子通道的纳米结构电极对于提高各种能源转化与储存装置(例如,超级电容器[53]、锂电池[54]、太阳能电池[55]、固态燃料电池[56]以及光电化学池[57])的性能是十分重要的。近些年来,超级电容器因具有比静电电容器更高的能量密度、比电池更高的功率密度和更长的循环寿命而引起人们广泛的关注。根据文献报道,有序的阵列结构直接生长于导电基底上非常有利于促进电化学能源储存[58]。目前,制备纳米阵列电极的主要方法有化学气相沉积、模板辅助的电沉积、水热法等[59-61]。尽管这方面已取得一系列进展,但发展一种简便且可控的无模板法来合成纳米阵列电极仍然是很有意义的。

在目前应用于超级电容器电极材料的过渡金属氧化物中,二氧化锰是最引人瞩目的,因其具有理想的电容特征和较低的成本[62,63]。特别地,最近报道的一些一维(例如,纳米线、纳米管、纳米棒等)二氧化锰纳米阵列具有比粉体材料更优异的电容行为[64]。可是,到目前为止,文献中很少报道二维纳米片阵列的合成方法及其超级电容性能。

在本章,在臭氧气氛和室温下,脉冲激光溅射锰金属靶,产生的等离子体羽流与臭氧分子反应生成二氧化锰纳米片阵列。当作为超级电容器电极时,该纳米片阵列电极具有优异的倍率性能、较高的比电容和良好的循环寿命。

## 3.2 实验部分

### 3.2.1 实验原料与仪器

实验过程中用到的原料如表3-1所示。

表3-1 实验原料

| 原料名称 | 生产厂家 | 备注 |
| --- | --- | --- |
| 锰 | Alfa Aesar | 纯度≥99.995% |
| 氧气 | 上海化工院青磊特种气体有限公司 | 纯度≥99.995% |
| 硫酸钠 | 天津市福晨化学试剂厂 | 分析纯 |

注:实验过程中使用的仪器参见第2章实验仪器。

### 3.2.2 材料制备

在脉冲激光沉积二氧化锰纳米片的过程中,利用臭氧代替氧气作为氧化剂,臭氧是由纯氧气直接放电并在液氮中冷冻而得到的。在激光沉积之前,先把体系的压强通过机械泵和分子泵抽至 $10^{-4}$ Pa,随后臭氧通过针型阀通入体系中,压强保持在 30 Pa。基底(镀金的镍片)与锰靶之间的距离是 3 cm。将激光聚焦于高纯锰靶材表面上,形成的等离子体羽流与臭氧分子反应,产物沿靶材表面法线方向飞出,碰到距离靶材约 3 cm 处的基底后,

# 第3章 脉冲激光沉积大面积二氧化锰纳米片阵列及其在超级电容器中的应用

在其表面沉积形成二氧化锰薄膜。在沉积过程中靶材通过马达不断旋转,以提高沉积速率和沉积薄膜的均匀性。为了对比,也制备了氧气气氛下(其他参数相同)的二氧化锰电极。

### 3.2.3 材料表征

1. 物理表征

实验中利用场发射扫描电子显微镜(FESEM;Philips XSEM30,Holland)和透射电镜(TEM;JEOL,JEM-2010,Japan)观测样品表面结构,通过X射线衍射(XRD:D/Max-2400;Cu 靶;$\lambda=1.541\,8\,\text{Å}$;管电压 40 kV;管电流 60 mA;扫描速度 5°/min),显微共聚焦拉曼光谱仪(514 nm laser with RM100)和X射线光电子能谱(PerkinElmer PHI 6 000C ECSA;Al KR/1 486.6 eV)对样品进行了结构分析。

2. 电化学表征

上述得到的二氧化锰阵列电极直接作为工作电极,在室温下采用三电极测试系统在 CHI660d 电化学工作站上进行。所有测试均在 1 M $Na_2SO_4$ 中进行,以铂片做对电极,饱和甘汞做参比电极。

## 3.3 结果与讨论

图 3-1 为脉冲激光沉积二氧化锰纳米片阵列的示意图。聚焦后的高能脉冲激光瞬间熔化锰靶,产生锰的等离子体羽流与臭氧分子反应生成二氧化锰,沉积在导电基底上,经过簇团的生长,最后形成纳米片阵列结构。图 3-2(a)为纳米片阵列的低倍扫描图片,由此可以看出,二氧化锰纳米片大面积地垂直生长在基底上,片与片相互交织形成三维多孔网络结构,非常有利于电解质离子的渗透。从高倍电镜图片可以观察到(图 3-2(b)),

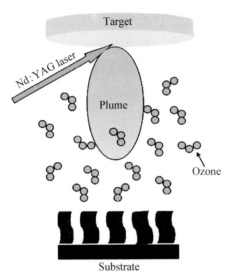

图 3-1 脉冲激光沉积二氧化锰纳米片阵列的示意图

单个纳米片的厚度只有 10 nm。纳米片的这种超薄结构也可以从透射电镜图片得以证实。如图 3-2(c)所示,单个的二氧化锰纳米片呈现出一些皱纹和弯曲的现象,类似于石墨烯片,因为二维结构在热力学上是不稳定的,通过弯曲可以降低表面张力。另外,由图 3-2(c)可知,纳米片的厚度约为 800 nm。X 射线衍射结果显示,二氧化锰纳米片为无定形结构,这与以前文献报道的结构一致[65]。为了进一步确认二氧化锰的结构,利用拉曼光谱和 X 射线光电子能谱进行测试。拉曼光谱结果显示,在 560 cm$^{-1}$ 和 646 cm$^{-1}$ 处,有 2 个二氧化锰的特征峰,分别归属于 Mn—O 键的平面伸缩振动和对称伸缩振动[66]。光电子能谱 Mn 2p 的信号,位于 644.0 eV(Mn 2p$_{3/2}$)和 655.6 eV(Mn 2p$_{1/2}$),与文献报道的二氧化锰的 XPS 结果一致[67,68]。

电容性能测试在 1 M Na$_2$SO$_4$ 溶液中进行。图 3-3(a)和图 3-3(b)为纳米片阵列在 2 mV·s$^{-1}$ 到 300 mV·s$^{-1}$ 时的循环伏安曲线。在此扫速范围内,循环伏安曲线的形状始终保持矩形,揭示了快速的充放电过程。由图 3-3(c)和图 3-3(d)看到,在每一个电流密度下的充放电曲线呈线性关系且保持对称,证明了良好的可逆性。在 1 A·g$^{-1}$ 时,比电容达 337 F·g$^{-1}$;在电流密度高达 100 A·g$^{-1}$ 时,电容保持率为 52%(相对于 1 A·g$^{-1}$)(图 3-3(e))。特别指出的是,在 100 A·g$^{-1}$ 时,一个完整的充放电过程仅需要 2.8 s。这种良好的倍率性能甚至超过文献报道的最优性能。例如,文献报道的二氧化锰纳米棒阵列电流密度从 3 A·g$^{-1}$ 到 30 A·g$^{-1}$ 时,电容保持

# 第3章 脉冲激光沉积大面积二氧化锰纳米片阵列及其在超级电容器中的应用

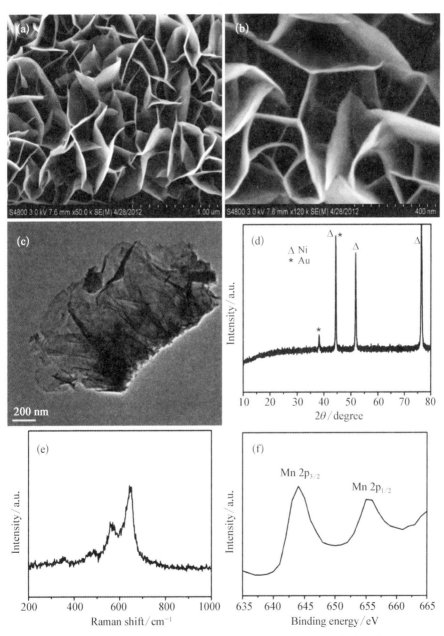

图3-2 二氧化锰纳米片阵列的扫描电镜图片(a,b),透射电镜图片(c), X射线衍射图(d),拉曼光谱(e),X射线光电子能谱(f)

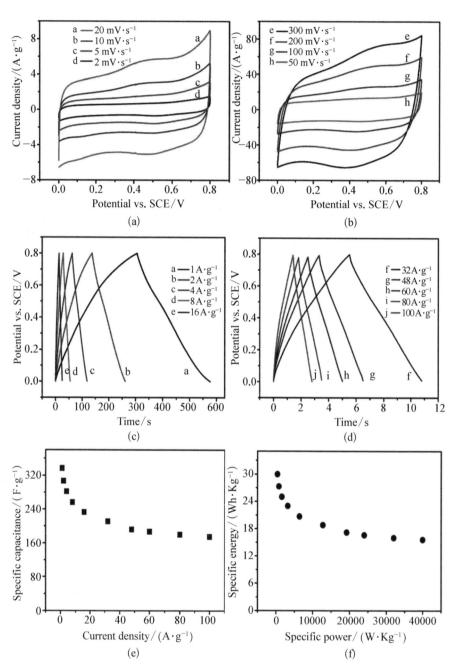

图3-3 二氧化锰纳米片阵列的电化学性质：(a,b)循环伏安曲线，(c,d)充放电曲线，(e)比电容值，(f)能量-功率图

# 第 3 章 脉冲激光沉积大面积二氧化锰纳米片阵列及其在超级电容器中的应用

率为 48%[64]。另外，$MnO_2/Zn_2SnO_4$ 电极的电流密度从 $1\ A\cdot g^{-1}$ 到 $40\ A\cdot g^{-1}$ 时[69]，电容保持率为 64%。能量密度和功率密度的关系如图 3-3(f) 所示[70]，最高能量密度在功率密度为 $400\ W\cdot kg^{-1}$ 时达到 $30\ Wh\cdot kg^{-1}$，最高功率密度为 $40\ kW\cdot kg^{-1}$。

为了进行对比，在氧气气氛下（其他参数相同）制备了二氧化锰电极，其电化学性质如图 3-4 所示。氧气气氛得到的二氧化锰薄膜电极在电流密度为 $1\ A\cdot g^{-1}$ 时，比电容达 $238\ F\cdot g^{-1}$；在电流密度为 $32\ A\cdot g^{-1}$ 时，电容保持率为 50%（相对于 $1\ A\cdot g^{-1}$）。这些电容值明显低于臭氧气氛下制备的二氧

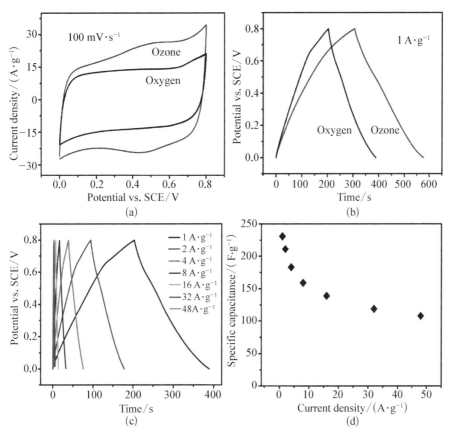

图 3-4 在氧气气氛下制备的二氧化锰电极的电化学性质：(a) 循环伏安曲线，(b,c) 充放电曲线，(d) 比电容值

化锰纳米片阵列。这也表明,高氧化性的臭氧更有利于制备高电容性能的电极[71]。

长的循环寿命对于超级电容器的实际应用是非常重要的,如图3-5所示,二氧化锰纳米片阵列在 300 mV·s$^{-1}$ 经过 6 000 次充放电循环后,电容值不但没有衰减,反而有所增加。此结果证明了纳米片电极良好的循环寿命。

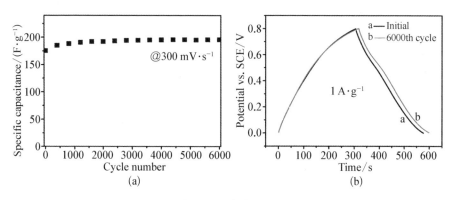

图3-5 二氧化锰纳米片阵列的循环性能

二氧化锰纳米片阵列电极的优异电容性能是与其结构有直接关系的。首先,激光溅射出来的粒子,历经反应后紧密结合于基底上,有效降低了集流体与活性物质的接触电阻;其次,高度有序且超薄的纳米片提供了有效的比表面积,有利于快速的氧化还原反应;最后,由于二氧化锰直接生长在基底上,不需要任何黏结剂,从而减少了活性材料的"死体积"。

## 3.4 本章小结

在室温下,利用臭氧为氧化剂,通过脉冲激光沉积法成功制备了二氧化锰纳米片阵列。该薄膜的多孔结构十分有利于电荷的储存,进而展现出良好的超级电容性能。另外,该结构也可应用于锂离子电池,氧还原反应等。

# 第 4 章

构建基于氧化镍/石墨烯泡沫和多孔氮掺杂碳纳米管的不对称超级电容器及其超高的倍率性能

## 4.1 引　　言

作为一种新型的能量储存装置,超级电容器因其比电池具有更高的功率密度和更好的循环寿命而适用于电动汽车和可再生能源体系[72-74]。可是,现有超级电容器的能量密度远低于二次电池,从而限制了其广泛的应用[75,76]。众所周知,能量密度与电位窗口的平方成正比,因此有机电解质的应用在某种程度上提高了超级电容器的能量密度[77-81]。但是,有机电解质本身价格昂贵且大部分有毒,不宜于商业化应用。最理想的方案就是在水溶液中实现较宽的电位窗口。近期,不对称超级电容器的出现成功解决了这一难题[82-85]。

迄今为止,一系列氧化还原性材料(例如金属氧化物/氢氧化物、导电聚合物)被应用于不对称超级电容器的正极材料[86]。其中,氧化镍因其具有高的理论比电容($2\,583\ \text{F}\cdot\text{g}^{-1}$)、高的化学/热稳定性、良好的可逆性、环境友好和低成本而倍受青睐[87-89]。例如,文献报道的以氧化镍为正极、多孔碳为负极的不对称电容器具有 11.6 Wh·kg$^{-1}$ 的能量密度[90]。为了进一步提高氧化镍的电容性能,一些文献把氧化镍与还原氧化石墨烯进行复合,其电容值有了一定的提高[91-96]。然而,还原氧化石墨烯本身的导电性和比表面积因其大量缺陷存在和石墨烯片之间的 π—π 作用而失去优势。最近,通过 CVD 制备的石墨烯泡沫成功避免了上述问题,因为该泡沫具有三维连续无缺陷和高度导电的石墨烯网络[97-99]。这种高质量的石墨烯泡沫为离子和电子的传输提供了高导电性的通道。因此,十分期待将氧化镍直接生长于石墨烯泡沫的电容性能。

就不对称超级电容器的负极材料而言,活性炭是最常用的材料,因其具有高比表面积和相对高的导电性。然而,活性炭的高倍率性能较差,例如,在无定型 Ni(OH)$_2$//活性炭电容器中[100],活性炭在电流密度为 4.8 A·g$^{-1}$ 时,比电容仅有 120 F·g$^{-1}$。这主要是由于活性炭自身缺乏有效的离子传输通道[101,102]。因此,设计一种三维多层次孔状的碳材料对于提高不对称超级电容器的性能也是十分关键的。

在本章,我们首先利用脉冲激光沉积法制备了氧化镍/石墨烯泡沫,作为不对称电容器的正极材料;其次,以聚吡咯纳米管为前驱体、KOH 为活化剂合成了功能化多孔氮掺杂的碳纳米管,作为不对称电容器的负极材料;最后,以 KOH 溶液为电解质构建了一种新型的不对称超级电容器(NiO/GF//HPNCNT)。

## 4.2 实验部分

### 4.2.1 实验原料与仪器

实验过程中用到的原料和仪器如表 4-1 和表 4-2 所示。

第4章 构建基于氧化镍/石墨烯泡沫和多孔氮掺杂碳纳米管的不对称超级电容器及其超高的倍率性能

表 4-1 实验原料

| 原料名称 | 生产厂家 | 备注 |
| --- | --- | --- |
| 镍 | Alfa Aesar | 纯度≥99.995% |
| 氧气 | 上海化工院青磊特种气体有限公司 | 纯度≥99.995% |
| 氢氧化钾 | 江苏强盛化学股份有限公司 | 分析纯 |
| 吡咯 | 上海辰谊科学仪器有限公司 | 化学纯 |
| 甲基橙 | 上海辰谊科学仪器有限公司 | 分析纯 |
| 无水乙醇 | 天津市福晨化学试剂厂 | 分析纯 |

表 4-2 实验仪器

| 仪器型号及名称 | 生产厂家 |
| --- | --- |
| CHI660d 电化学工作站 | 上海辰华仪器公司 |
| CT2001A LAND 电池测试系统 | 武汉市金诺电子有限公司 |
| BT25S 电子天平 | 赛多利斯科学仪器(北京)有限公司 |
| SRJX-4-13 高温箱式电阻炉 | 北京市永光明医疗仪器厂 |
| SHB-IV 循环水式多用真空泵 | 郑州长城科工贸有限公司 |
| SK3300LH 超声波清洗器 | 上海科导超声仪器有限公司 |
| 101A-1 型电热鼓风干燥箱 | 上海实验仪器厂有限公司 |
| 90-1 型恒温磁力搅拌器 | 上海沪西分析仪器厂 |
| DF-101B 集热式恒温磁力搅拌器 | 浙江省乐清市乐成电器厂 |
| 脉冲激光沉积系统 | 同第2章 |

## 4.2.2 材料的制备

**1. 氧化镍/石墨烯泡沫的合成**

根据以前文献报道,石墨烯泡沫通过 CVD 制备[97]。以石墨烯泡沫为基底,氧化镍的脉冲激光沉积过程如图 4-1 所示,同第 3 章参数,其中,镍靶代替锰靶。

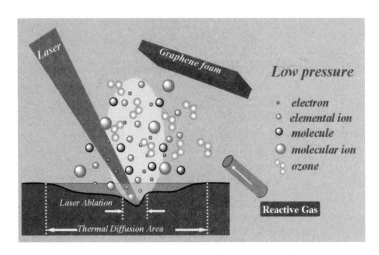

图 4-1　脉冲激光沉积制备氧化镍/石墨烯泡沫的示意图

2. 多孔氮掺杂碳纳米管的合成

制备过程分两步进行：第一步，根据以前文献合成聚吡咯纳米管[103,104]；第二步，将聚吡咯纳米管与氢氧化钾以 1∶4 质量比混合，在 850℃氮气气氛下退火 1 h，等降到室温后，用 1 mol HCl 洗涤直至 pH 变为中性，60℃真空烘干。

### 4.2.3　材料表征

1. 物理表征

实验中利用场发射扫描电子显微镜（FESEM；Philips XSEM30，Holland）和透射电镜（TEM；JEOL，JEM‑2010，Japan）观测样品表面结构，显微共聚焦拉曼光谱仪（514 nm laser with RM100）和 X 射线光电子能谱（PerkinElmer PHI 6 000C ECSA；Al KR/1 486.6 eV）对样品进行了结构分析，氮气吸脱附曲线通过 Micromeritics Tristar 3 000 analyzer 在液氮温度下测试。

2. 电化学表征

上述得到的氧化镍薄膜直接作为工作电极，在室温下采用三电极测试系统在 CHI660d 电化学工作站上进行。所有测试均在 1 M KOH 中进行，以铂片做对电极，饱和甘汞做参比电极。

工作电极的制备：将多孔氮掺杂碳纳米管与聚四氟乙烯按 9∶1 的质量比混合制作电极，先充分研磨后再加入聚四氟乙烯的乳液使其混合均匀后压在碳纤维纸（0.1 mm 厚度）上作为测试电极。氧化镍/石墨烯泡沫直接作为工作电极。

电化学性能测试：在室温下，首先采用三电极体系，在 CHI660d 电化学工作站上测试氧化镍/石墨烯泡沫和多孔氮掺杂碳纳米管分别在正区和负区单电极的电容性能。所有测试均在 1 M KOH 中进行，以铂片做对电极，Hg/HgO 做参比电极。

不对称超级电容器的组装：在两电极体系下，以氧化镍/石墨烯泡沫为正极，多孔氮掺杂碳纳米管为负极，聚丙烯膜为隔膜，1 M KOH 为电解质，组装成三明治结构，活性物质氧化镍和多孔碳纳米管的质量比为 1∶2。

## 4.3 结果与讨论

### 4.3.1 正极材料

图 4-1 是脉冲激光沉积法制备氧化镍/石墨烯泡沫的示意图，其原理和过程与第 2 章中表述的相同[105]。图 4-2(a)～(d)为石墨烯泡沫的扫描电镜图片。从低倍图片可以看出，石墨烯泡沫完全复制了泡沫镍的完整连续的大孔结构；在高倍电镜下，可以清晰地看到石墨烯表面的"皱纹"，这是由于镍和碳的热膨胀系数的不同而造成的。石墨烯泡沫经过沉积氧化镍后，照片颜色变得更深。从扫描电镜图片可以看出，氧化镍纳米粒子均匀地沉积在石墨烯表面。透射电镜图片如图 4-3 所示，可以清晰地观察到大

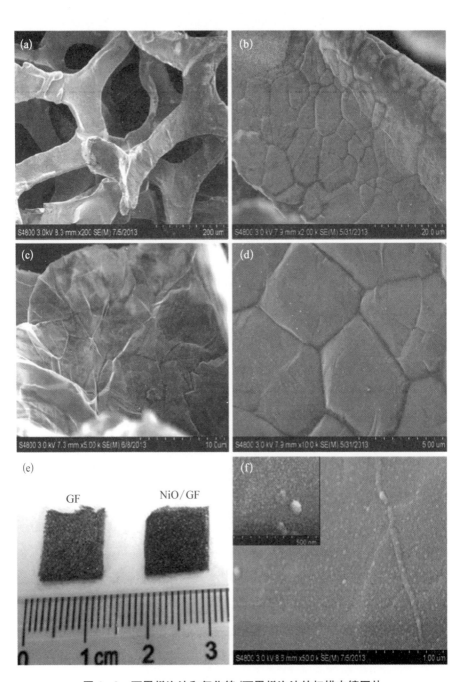

图4-2 石墨烯泡沫和氧化镍/石墨烯泡沫的扫描电镜图片

第4章 构建基于氧化镍/石墨烯泡沫和多孔氮掺杂碳纳米管的不对称超级电容器及其超高的倍率性能

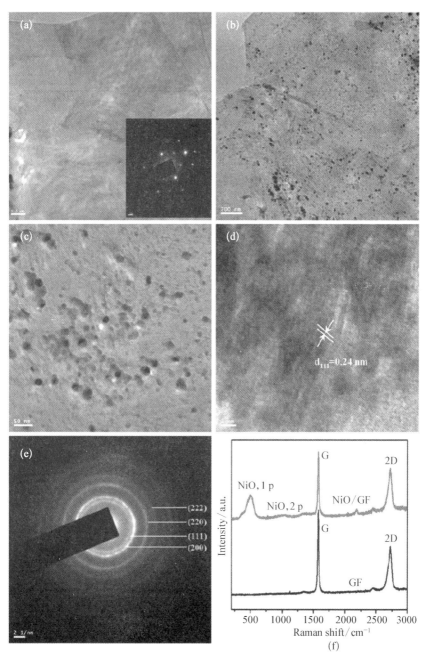

图4-3 石墨烯泡沫(a)和氧化镍/石墨烯泡沫(b-d)的
透射电镜图片、(e)选区衍射和(f)拉曼光谱

尺寸的石墨烯片,选区衍射显示了单晶结构[106]。从氧化镍/石墨烯泡沫的图片可以看到,约 20 nm 大小的氧化镍粒子分散在石墨烯片上。高倍透镜图片可以清晰地看到(111)晶面的晶格条纹。选区衍射环表明了氧化镍的多晶结构。拉曼光谱用来进一步分析样品的微结构,如图 4-3(f)所示,石墨烯泡沫没有出现 D 峰,说明石墨烯的高质量和高导电性。由 G 峰和 2D 峰比例可以看到,该石墨烯泡沫是由几层石墨烯组成[107,108]。与原始的石墨烯泡沫相比,氧化镍/石墨烯泡沫的拉曼光谱除了石墨烯的峰之外,在 510 $cm^{-1}$ 和 1 040 $cm^{-1}$ 处的峰对应于氧化镍的 1 p 和 2 p 散射[109,110]。这些拉曼信号有力地证明了氧化镍成功生长在石墨烯泡沫上。三维大孔高导电性的石墨烯泡沫与高度分散的氧化镍纳米粒子的紧密结合将有利于电容性能的提高。

氧化镍/石墨烯泡沫在三电极体系下的赝电容行为如图 4-4 所示。图 4-4(a)为石墨烯泡沫和氧化镍/石墨烯泡沫在 5 $mV·s^{-1}$ 时的循环伏安曲线。相比于氧化镍/石墨烯泡沫,石墨烯泡沫的循环伏安曲线几乎是一条直线,表明石墨烯泡沫没有电容贡献,仅作为集流体,而电容的贡献主要是来自氧化镍。氧化镍/石墨烯泡沫的循环伏安曲线有一对明显的氧化还原峰,对应的电极反应为:$NiO+OH^- \longleftrightarrow NiOOH+e^-$,其峰位差仅为 100 mV,远低于文献中的 NiO/Ni 复合物[87]、NiO 纳米柱[111]和 NiO 纳米带[112],表明了氧化镍/石墨烯泡沫的良好可逆性。比电容值可以通过充放电曲线来计算,在电流密度为 1 $A·g^{-1}$ 时,比电容达 1 225 $F·g^{-1}$。在电流密度高达 100 $A·g^{-1}$ 时,电容值仍达 827 $F·g^{-1}$(相对于 1 $A·g^{-1}$ 电容保持率为 68%)。如此之高的倍率性能是目前文献报道中最好的结果[87,110-112]。此外,该氧化镍薄膜电极具有良好的循环寿命,在 1 000 个充放电循环后,电容保持 89%。这些好的电容性能与氧化镍/石墨烯泡沫的高导电性紧密相关。因此,氧化镍/石墨烯泡沫也可以作为不对称超级电容器的正极材料。

第4章 构建基于氧化镍/石墨烯泡沫和多孔氮掺杂碳纳米管的不对称超级电容器及其超高的倍率性能

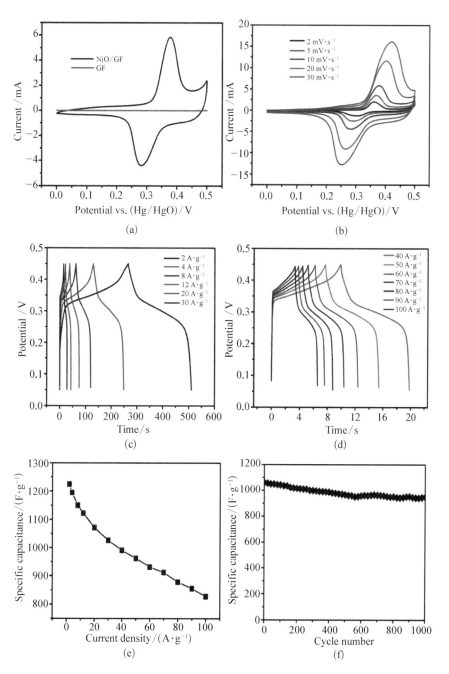

图4-4 氧化镍/石墨烯泡沫的循环伏安曲线(a,b)、充放电曲线(c,d)、比电容值(e)和循环性能(f)

### 4.3.2 负极材料

负极材料对于提高不对称超级电容器的性能也是很关键的,然而,大多数有关不对称电容器的报道主要集中于优化正极材料,而忽略了负极材料。在本章,我们以聚吡咯纳米管为前驱体,通过化学活化的方式制备了氮掺杂的层次孔状的碳纳米管。图4-5为碳纳米管的电镜图,其外径为190 nm,内径为129 nm。经过活化后的碳纳米管的外径和内径分别为130 nm和70 nm(图4-6)。活化后直径的变小是由于在活化过程中质量的损失造成的[113]。拉曼光谱证明了碳纳米管的形成(图4-6(f))。另外,

图4-5 聚吡咯纳米管的扫描(a,b)和透射(c,d)电镜图片

第4章 构建基于氧化镍/石墨烯泡沫和多孔氮掺杂碳纳米管的不对称超级电容器及其超高的倍率性能

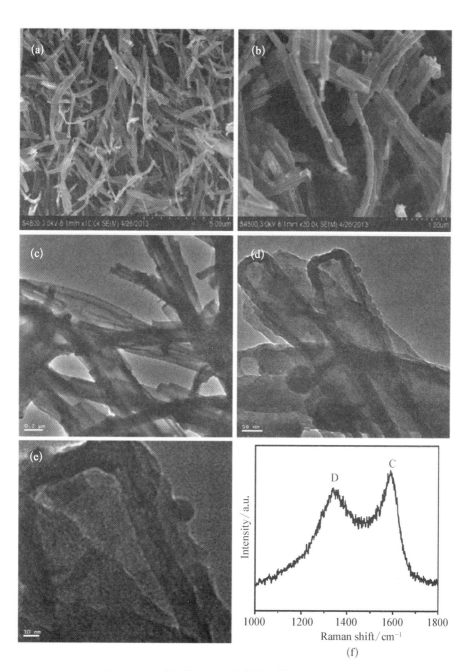

图4-6 多孔氮掺杂碳纳米管的扫描电镜图片(a,b)、
透射电镜图片(c-e)和拉曼光谱(f)

根据氮气吸附曲线(图 4-7(a)),碳纳米管的比表面积达到 2 080 $m^2 \cdot g^{-1}$;根据孔径分布图(图 4-7(b)),多孔碳纳米管在 3.71 nm 和 1.89 nm 附近有明显的介孔和微孔,总的孔容达到 1.23 $m^3 \cdot g^{-1}$。经过元素分析,多孔碳纳米管中的氮和氧的含量分别为 2.8 wt% 和 12.4 wt%。XPS 测试表明(图 4-7(c)),在 284.8 eV,532.0 eV,和 400.0 eV 处的峰,分别属于 C 1 s,O 1 s,N 1 s。将 N 1 s 进行分峰后,在 401.5,400.3 和 398.3 eV 位置,归为四元型、吡咯型和吡啶型氮[114]。

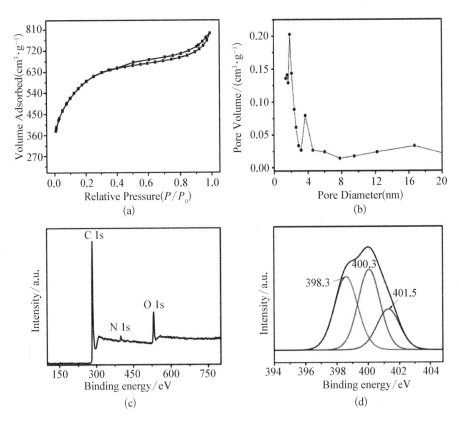

图 4-7 多孔氮掺杂碳纳米管的氮气吸脱附曲线(a)、孔径分布(b)和 XPS 数据(c,d)

图 4-8(a)和图 4-8(b)是多孔碳纳米管在不同扫速下的循环伏安曲线,在 500 mV·$s^{-1}$ 时,仍保持矩形形状,表明具有很理想的电容行为。通

第4章 构建基于氧化镍/石墨烯泡沫和多孔氮掺杂碳纳米管的不对称超级电容器及其超高的倍率性能

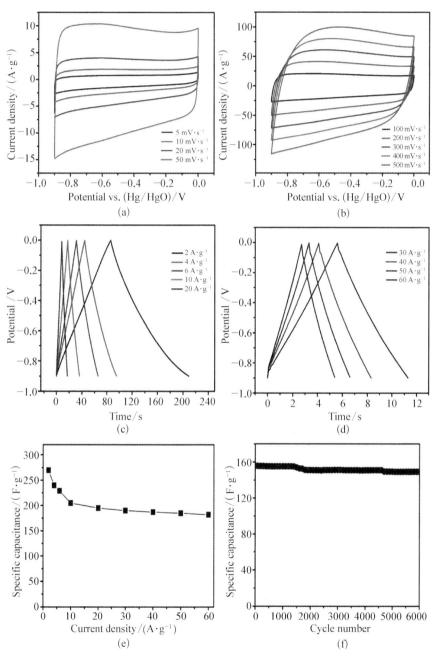

图4-8 多孔氮掺杂碳纳米管的循环伏安曲线(a,b)、
充放电曲线(c,d)、比电容值(e)和循环性能(f)

过充放电曲线计算出比电容,在电流密度为 2 A·g$^{-1}$ 时达到 270 F·g$^{-1}$;在电流密度为 60 A·g$^{-1}$ 时达到 182 F·g$^{-1}$,说明具有良好的倍率性能。在扫速为 500 mV·s$^{-1}$ 经过 6 000 次循环后,比电容仅衰减了 4%,揭示了优异的循环性能。碳纳米管的这些优异电容性能主要归因于① 大量的微孔、介孔提供了超高的比表面积,使其具有高比电容;② 纳米管结构提供了有效的离子通道;③ 高含量的氮元素掺杂增加了导电性和可润湿性。多孔氮掺杂的碳纳米管的这些高电容性能使其作为不对称超级电容器负极材料是十分有益的。

### 4.3.3 不对称超级电容器

基于氧化镍/石墨烯泡沫正极(0.0~0.5 V)和多孔碳纳米管(−0.9~0.0 V)的电位窗口,不对称电容器的电位窗口可以扩展为 1.4 V。根据正负极电荷的平衡关系[115]:$q^+ = q^-$ 和 $q = C \times \Delta V \times m$,正负极活性物质的质量最优比为 0.5。

从不对称电容器的循环伏安曲线可以看出(图 4-9(a)),由于正极氧化镍的电容贡献,循环伏安图不是呈现一个矩形,而是具有明显的氧化还原峰。从充放电曲线可以观察到,在电流密度为 60 A·g$^{-1}$ 时,放电曲线仍没有出现明显的电位降,表明不对称电容器具有很低的电阻。同时,所有的充放电曲线非常对称,说明该不对称电容器具有很高的可逆性和库伦效率(图 4-9(b)和图 4-9(c))。不同电流密度下的比电容值如图 4-9(d)所示,在电流密度为 1 A·g$^{-1}$ 时达到 116 F·g$^{-1}$;在电流密度为 60 A·g$^{-1}$ 时仍保持 61 F·g$^{-1}$,证明了良好的倍率性能。在电流密度为 10 A·g$^{-1}$ 经过 2 000 次循环后,比电容仅衰减了 6%,表明具有优异的循环性能,同时也要优于其他文献报道的不对称电容器的循环稳定性,例如 Ni(OH)$_2$//AC(1 000 次循环 82% 保持率)[116],graphene/MnO$_2$//graphene(1 000 次循环 79% 保持率)[115],NiCo$_2$O$_4$-rGO//AC(2 500 次循环 83% 保持率)[117],Ni-

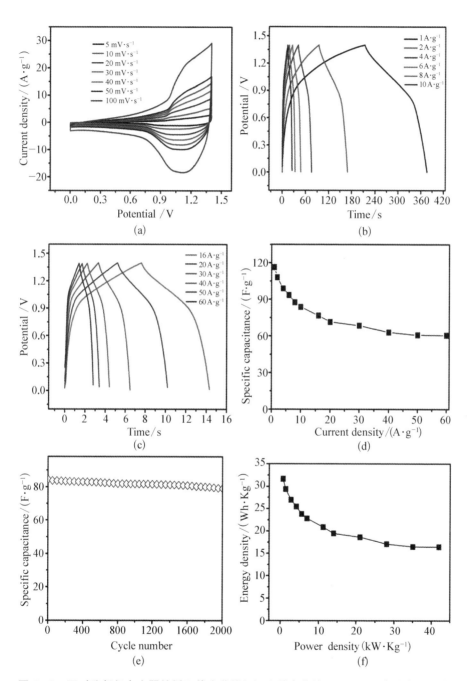

图4-9 不对称超级电容器的循环伏安曲线(a)、充放电曲线(b,c)、不同电流密度下的比电容值(d)、循环性能(e)和能量-功率关系图(f)

Co oxide//AC(2 000次循环85%保持率)[118]。不对称电容器的这些优异电容性能归因于正负极材料之间的正协同效应。

能量密度和功率密度是描述超级电容器性能的两个关键因素。图4-9(f)是不对称超级电容器的能量-功率关系图。如图所示，当功率密度从 0.7 kW·kg$^{-1}$ 增加到 42 kW·kg$^{-1}$，能量密度则从 32 Wh·kg$^{-1}$ 降到 17 Wh·kg$^{-1}$。这些性能远高于文献报道的一些不对称超级电容器的能量和功率值，例如 Ni(OH)$_2$/graphene//porous graphene（13.5 Wh·kg$^{-1}$—15.2 kW·kg$^{-1}$）[86]，graphene/Ni(OH)$_2$//graphene/RuO$_2$（14 Wh·kg$^{-1}$—21 kW·kg$^{-1}$）[119]，graphene/MnO$_2$//graphene（7 Wh·kg$^{-1}$—5 kW·kg$^{-1}$）[115]，CoO@PPy nanowire array//AC（11.8 Wh·kg$^{-1}$—5.5 kW·kg$^{-1}$）[120]，graphene-NiCo$_2$O$_4$//AC（7.6 Wh·kg$^{-1}$—5.6 kW·kg$^{-1}$）[121]，graphene/MnO$_2$//ACN（8.2 Wh·kg$^{-1}$—16.5 kW·kg$^{-1}$）[122]和 RGO-RuO$_2$//RGO-PANi（6.8 Wh·kg$^{-1}$—49.8 kW·kg$^{-1}$）[123]。另外特别要说明的是，不对称超级电容器所具有的最高功率密度为 42 kW·kg$^{-1}$，意味着一个完整的充放电只需要 2.8 s，完全符合下一代电动汽车的功率需求（PNGV：Partnership for a New Generation of Vehicles）[124-127]。这也同时意味着，所制备的氧化镍/石墨烯泡沫//多孔氮掺杂碳纳米管不对称电容器在新能源领域具有潜在的应用前景。

## 4.4 本章小结

本章基于氧化镍/石墨烯泡沫为正极、多孔氮掺杂的多孔氮掺杂的碳纳米管为负极在 KOH 溶液中构建了一种新型的具有超高倍率的不对称超级电容器。所制备的不对称超级电容器在 1.4 V 电位窗口下具有很高的能量密度、功率密度和循环稳定性。特别是在超高的功率密度（42 kW·kg$^{-1}$，一个完整的充放电仅需要 2.8 s）下，能量密度仍能达到 17 Wh·kg$^{-1}$，优于文献中报道的钴镍锰基的不对称电容器。

# 第5章
# 控制生长钴酸镍纳米线和纳米片在碳布上及其不同的赝电容行为

## 5.1 引　言

相比于单组分氧化物,二元尖晶石结构的钴酸物($MCo_2O_4$;M=Mn,Ni,Zn,Cu,Mg等)具有更丰富的氧化还原活性[128-130]。例如,$Co_3O_4$晶格中的部分钴被锰取代之后,对氧还原的催化活性明显提高[131,132]。另外,$NiCo_2O_4$具有比单个的NiO和$Co_3O_4$高两个数量级的导电性[133]。Lu及其合作中首次报道了$NiCo_2O_4$凝胶在超级电容器中的应用[134]。此后,有一系列特定纳米结构的$NiCo_2O_4$应用于电化学能量储存。根据电极的结构,这些关于$NiCo_2O_4$的工作可以分为两类:① 需要黏结剂的;② 不需要黏结剂的。对于前者,首先通过各种合成方法(例如球磨法[135]、共沉淀法[136-139]、模板法[140]、水热法[129,133,141,142]、溶胶凝胶法[134,143]等)制备$NiCo_2O_4$,再用传统的涂浆法制备工作电极。尽管这类工作电极取得了一系列进展,但是绝缘性黏结剂的使用增加了电极的电阻和一些"死体积"。对于后者,$NiCo_2O_4$纳米结构直接生长在导电基底上,直接作为工作电极,这样就避免了黏结剂的使用。$NiCo_2O_4$纳米针[144]、纳米线[145]和纳米

## 第5章　控制生长钴酸镍纳米线和纳米片在碳布上及其不同的赝电容行为

片[146,147]沉积在泡沫镍基底上具有超高的比电容。然而,也有文献报道,利用泡沫镍作为集流体,在碱性电解质中会给电容计算带来很大的误差[148-150]。此外,Gupta 等人在不锈钢基底上电沉积 $NiCo_2O_4$ 薄膜[151],表现出极高的倍率性能。然而,非柔性本质不利于其在弯曲或折叠情况下的实际应用[152]。

碳布作为一种柔性且低成本的基底,具有一些独特的优势,例如高强度、高导电性和强抗腐蚀性能。碳布所具有的这些特征使其在场发射[153]、锂电[154]、染料敏化太阳能电池[155]以及光检测器[156]等方面具有潜在的应用前景。考虑到类似的电极结构需求,碳布有望作为集流体在超级电容器方面具有突出的优势[157]。另外,一维和二维纳米结构已广泛应用于储能领域,但是,至今还没有关于在碳布基底上生长一维和二维 $NiCo_2O_4$ 在超级电容器中的应用。

在本章,以碳布为基底,均匀生长了 $NiCo_2O_4$ 的纳米线和纳米片两种结构,并考察哪种结构更有利于电化学能量储存,同时揭示结构与性能的关系。

## 5.2　实　验　部　分

### 5.2.1　实验原料与仪器

实验过程中用到的原料和仪器如表 5-1 所示。

表 5-1　实 验 原 料

| 原料名称 | 生 产 厂 家 | 备　注 |
| --- | --- | --- |
| 尿素 | 上海辰谊科学仪器有限公司 | 分析纯 |
| 氟化铵 | 上海辰谊科学仪器有限公司 | 分析纯 |
| 硝酸钴 | 上海辰谊科学仪器有限公司 | 分析纯 |

续 表

| 原料名称 | 生产厂家 | 备 注 |
|---|---|---|
| 硝酸镍 | 上海辰谊科学仪器有限公司 | 分析纯 |
| 氢氧化钾 | 江苏强盛化学股份有限公司 | 分析纯 |
| 无水乙醇 | 天津市福晨化学试剂厂 | 分析纯 |

注：制备过程中用到的仪器参见第 4 章表 4-2 实验仪器。

### 5.2.2 材料的制备

在沉积之前，碳布在 1 M $H_2SO_4$、去离子水、乙醇中分别超声 15 min，烘干后，置于水热反应釜中。纳米线的制备过程如下：1 mmol $Ni(NO_3)_2 \cdot 6H_2O$，2 mmol $Co(NO_3)_2 \cdot 6H_2O$，2 mmol $NH_4F$，6 mmol $CO(NH_2)_2$ 通过搅拌溶解在 80 mL 乙醇和水(体积比为 1∶1)中，将该溶液转移至上述放有碳布的反应釜中，在 95℃加热 10 h。等冷却至室温后，取出沉积了 $NiCo_2O_4$ 的碳布，洗涤、干燥之后，在 350℃退火 6 h。$NiCo_2O_4$ 纳米线在碳布上的负载量为 0.52 mg·cm$^{-2}$。

纳米片的制备过程如下：1 mmol $Ni(NO_3)_2 \cdot 6H_2O$，2 mmol $Co(NO_3)_2 \cdot 6H_2O$，1 g 十六烷基三甲基溴化铵溶解在 80 mL 甲醇和水(体积比为 5∶1)中，将该溶液转移至上述放有碳布的反应釜中，在 180℃加热 12 h。等冷却至室温后，取出沉积了 $NiCo_2O_4$ 的碳布，洗涤、干燥之后，在 350℃退火 6 h。$NiCo_2O_4$ 纳米片在碳布上的负载量为 0.6 mg·cm$^{-2}$。

### 5.2.3 材料表征

**1. 物理表征**

实验中利用场发射扫描电子显微镜(FESEM；Philips XSEM30，Holland)和透射电镜(TEM；JEOL，JEM-2010，Japan)观测样品表面结构。氮气吸脱附曲线通过 Micromeritics Tristar 3000 analyzer 在液氮温度下测试。

## 第5章 控制生长钴酸镍纳米线和纳米片在碳布上及其不同的赝电容行为

**2. 电化学表征**

上述得到的 $NiCo_2O_4$/碳布直接作为工作电极,在室温下采用三电极测试系统在 CHI660d 电化学工作站上进行。所有测试均在 1 mol KOH 溶液中进行,以铂片做对电极,饱和甘汞做参比电极。

## 5.3 结果与讨论

### 5.3.1 $NiCo_2O_4$ 纳米线和纳米片的合成过程

$NiCo_2O_4$ 纳米线的反应过程如下[158-160]:

$$Ni^{2+} + 2Co^{2+} + 3xF^- \longrightarrow [NiCo_2F_{3x}]^{3(x-2)-} \tag{5-1}$$

$$CO(NH_2)_2 + H_2O \longrightarrow 2NH_3 + CO_2 \tag{5-2}$$

$$CO_2 + H_2O \longrightarrow CO_3^{2-} + 2H^+ \tag{5-3}$$

$$NH_3 \cdot H_2O \longrightarrow NH_4^+ + OH^- \tag{5-4}$$

$$[NiCo_2F_{3x}]^{3(x-2)-} + 1.5(2-y)CO_3^{2-} + 3yOH^- + nH_2O \longrightarrow$$
$$NiCo_2(OH)_{3y}(CO_3)_{1.5(2-y)} \cdot nH_2O + 3xF^- \tag{5-5}$$

$$2NiCo_2(OH)_{3y}(CO_3)_{1.5(2-y)} \cdot nH_2O + O_2 \longrightarrow$$
$$2NiCo_2O_4 + (3y+2n)H_2O + 3(2-y)CO_2 \tag{5-6}$$

$NiCo_2O_4$ 纳米片的反应过程如下[150]:

$$CH_3OH + H_2O \longrightarrow CH_3OH_2^+ + OH^- \tag{5-7}$$

$$xNi^{2+} + 2xCo^{2+} + 6xOH^- \longrightarrow Ni_xCo_{2x}(OH)_{6x} \tag{5-8}$$

$$2Ni_xCo_{2x}(OH)_{6x} + xO_2 \longrightarrow 2xNiCo_2O_4 + 6xH_2O \tag{5-9}$$

### 5.3.2 材料表征

NiCo$_2$O$_4$ 纳米线的扫描电镜图片如图 5-1 所示,从低倍图片可以看到,每一根碳纤维表面都均匀生长了一层纳米线。从高倍图片可以观察到,纳米线呈针状特征,垂直于碳纤维而耸立。NiCo$_2$O$_4$ 纳米线的透射电镜图片如图 5-2 所示,纳米线的直径由靠近碳布的底端至顶端的变化是 10～40 nm,而且纳米线表面分布着无数的介孔。由高倍透镜图片可知,层间距为 0.47 nm 的晶格条纹对应于尖晶石钴酸镍的(111)晶面。另外,从选区衍射环可以判断,NiCo$_2$O$_4$ 纳米线为多晶结构,所有的衍射环完全与尖

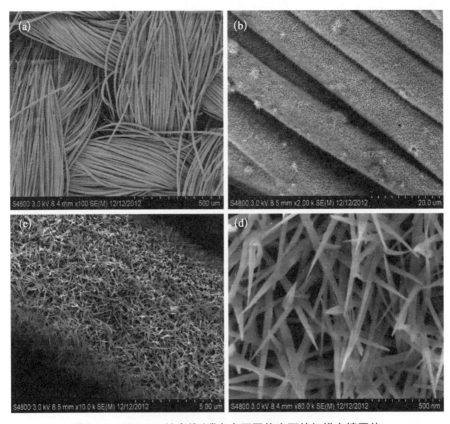

图 5-1　NiCo$_2$O$_4$ 纳米线/碳布在不同倍率下的扫描电镜图片

# 第 5 章  控制生长钴酸镍纳米线和纳米片在碳布上及其不同的赝电容行为

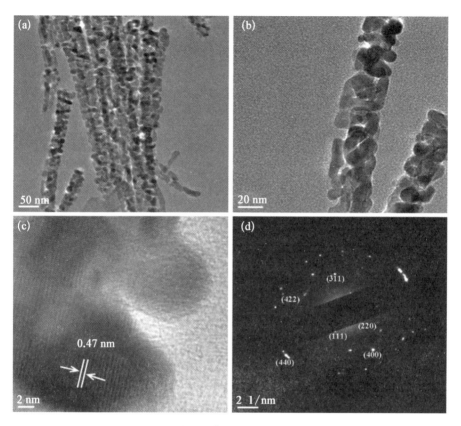

图 5‑2  $NiCo_2O_4$ 纳米线/碳布在不同倍率下的透射电镜图片

晶石 $NiCo_2O_4$ 晶体结构的晶面相对应(JCPDS No. 20‑0781，$a=b=c=0.811$ nm)。

$NiCo_2O_4$ 纳米片的扫描电镜图片如图 5‑3 所示，从低倍图片可以看到，每一根碳纤维表面都均匀生长了一层纳米片。从高倍图片可以观察到，纳米片垂直于碳纤维而生长。$NiCo_2O_4$ 纳米片的透射电镜图片如图 5‑4 所示，可以清楚地看到，纳米片是由无数的纳米粒子(大约 20 nm)排列而成，而且纳米粒子之间又有一定的空隙存在。在高倍透镜图片中，层间距为 0.20 nm 的晶格条纹对应于尖晶石钴酸镍的(400)晶面。另外，从选

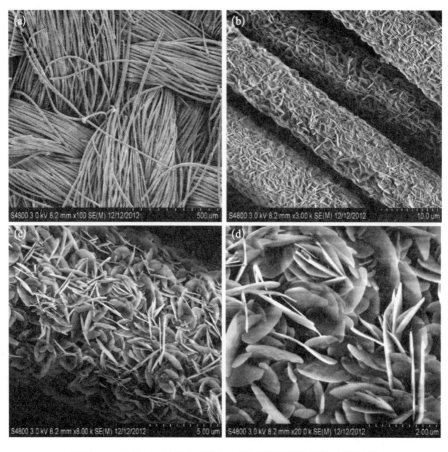

图 5-3 NiCo$_2$O$_4$ 纳米片/碳布在不同倍率下的扫描电镜图片

区衍射中的衍射环完全与尖晶石 NiCo$_2$O$_4$ 晶体结构的晶面相对应(JCPDS No. 20-0781，$a=b=c=0.811$ nm)。

为了进一步揭示纳米线和纳米片的微结构，进行氮气吸脱附测试，如图 5-5 所示。结果显示，纳米线和纳米片的比表面积分别是 79.34 m$^2$·g$^{-1}$ 和 28.22 m$^2$·g$^{-1}$。另外，纳米线在 4.86 nm 处有分布均匀的介孔，而纳米片的孔分布在 3.04，4.97，12.46 和 26.44 nm[161-163]。纳米线和纳米片的孔容分别为 0.26 cm$^3$·g$^{-1}$ 和 0.11 cm$^3$·g$^{-1}$。

第5章 控制生长钴酸镍纳米线和纳米片在碳布上及其不同的赝电容行为

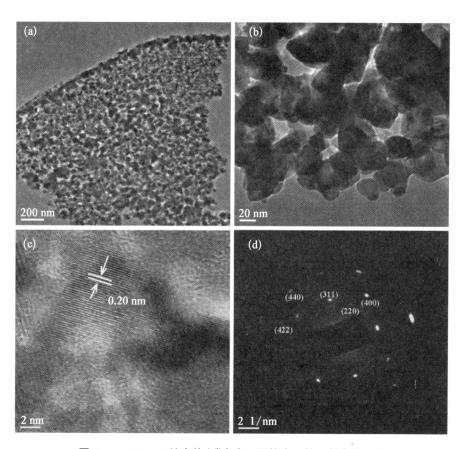

图5-4 NiCo$_2$O$_4$纳米片/碳布在不同倍率下的透射电镜图片

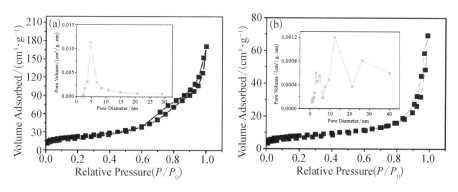

图5-5 NiCo$_2$O$_4$纳米线(a)和纳米片(b)氮气吸脱附曲线和孔径分布

### 5.3.3 电化学表征

图 5-6(a)和图 5-6(b)分别是 $NiCo_2O_4$ 纳米线和纳米片的循环伏安曲线,从曲线的形状来看,纳米片具有比纳米线更明显的氧化还原峰,但是纳米线具有比纳米片更大的积分面积,对应的电极反应是[164-167]:

$$NiCo_2O_4 + OH^- + H_2O \longleftrightarrow NiOOH + 2CoOOH + e^- \quad (5-10)$$

$$CoOOH + OH^- \longleftrightarrow CoO_2 + H_2O + e^- \quad (5-11)$$

从充放电曲线可以计算出比电容(图 5-6(c)和(d)),$NiCo_2O_4$ 纳米线的比电容在电流密度为 1,2,4,6,8,10 A·g$^{-1}$ 时分别为 245,239,228,217,203,191 F·g$^{-1}$。相比之下,$NiCo_2O_4$ 纳米片在电流密度为 1,2,4,6,8,

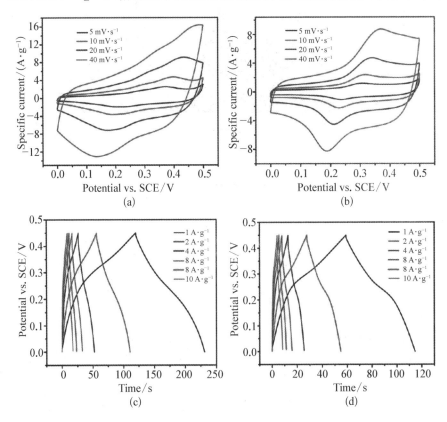

第5章 控制生长钴酸镍纳米线和纳米片在碳布上及其不同的赝电容行为

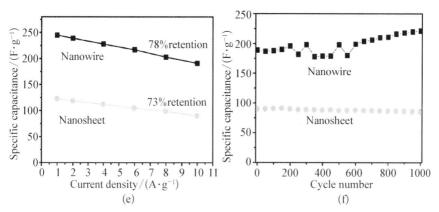

图 5-6 NiCo$_2$O$_4$ 纳米线(a,c)和纳米片(b,d)循环伏安曲线、充放电曲线、(e) 不同电流密度下的比电容值和(f) 循环性能

10 A·g$^{-1}$ 时,比电容分别是 123,119,112,105,99,90 F·g$^{-1}$。在电流密度为 10 A·g$^{-1}$ 下测试了循环性能,经过 1 000 次循环后,相比于初始电容,纳米线的比电容不但没有衰减反而有所增加,而纳米片的比电容衰减了 5%。由这些测试结果可以看出,在相同负载量的情况下,纳米线具有比纳米片更好的比电容、倍率特性和循环性能。这主要归因于:① 纳米线具有比纳米片更大的比表面积和孔容,从而更有利于离子的传输和电子的传导;② 从循环伏安曲线的形状可以推测,纳米线所具有的高电容性能可能是由于双电层电容和赝电容的共同贡献[168,169]。

## 5.4 本章小结

利用水热法在碳布基底上控制合成了介孔 NiCo$_2$O$_4$ 纳米线和纳米片。在相同负载量的情况下,纳米线和纳米片呈现出了截然不同的电容性能。相比于纳米片形貌,纳米线形貌具有更高的比电容值、倍率性能和循环性能。这种不同形貌之间的对比揭示了电化学能量储存中的"过程—结构—性质"的关系。

# 第6章
# 剪切多壁碳纳米管为弯曲石墨烯纳米片及其增强的电容性能

## 6.1 引 言

作为一种新型的能量储存装置,超级电容器因其优异的循环寿命、高功率密度和低成本而引起了科研人员的广泛关注[170-172]。电极材料是超级电容器的核心组成部分。近些年来,碳纳米管(CNTs)因为具有独特的纳米结构、高导电性和化学稳定性而被认为是良好的超级电容器电极材料[173]。然而,一般所制备的碳纳米管的长度在微米级且易于相互缠绕,非常不利于电解质离子在碳纳米管内传输,而导致较低的比电容($15 \sim 200 \text{ F} \cdot \text{g}^{-1}$)[174-176]。为了改善这种情况,文献报道了很多修饰碳纳米管的工作。

这些工作主要是通过横向剪切或纵向打开碳纳米管来进行的。有很多方法用来把碳纳米管从微米级切断成为纳米级,这些方法主要有球磨法[177]、低温粉碎[178]、固态反应[179,180]、选择性刻蚀[181]、氟化[182]、化学氧化[183]和电化学处理等[184]。通过横向切断后的碳纳米管在锂电[185]和太阳能电池[186]应用方面具有比原始碳纳米管更好的性能。另外,也可以通过等离子体刻蚀[187]、氧化处理[188,189]、电化学剥离[190]、钾蒸气处理[191]、催化氢化[192]和液氮热膨胀等方法将碳纳米管进行纵向打开形成石墨烯纳米带[193]。通常,纵向打开后的碳纳米管具有更高的氧化还原活性[194-196]。尽管已取得这些进展,但是关于进一步横向剪切和纵向打开碳纳米管的工作却很少报道。

在本章,通过修饰的 Hummers 法[197]将多壁碳纳米管沿着横向和纵向剪切,直接转化成弯曲的石墨烯片(Curved graphene nanosheets,CGN)。所得到的 CGN 保持了部分管状结构,但同时具有石墨烯的片状结构,当应用于超级电容器电极材料时,具有很高的电容性能。

## 6.2 实 验 部 分

### 6.2.1 实验原料与仪器

实验过程中用到的原料和仪器如表 6-1 所示。

表 6-1 实 验 原 料

| 原料名称 | 生 产 厂 家 | 备 注 |
| --- | --- | --- |
| 碳纳米管 | 深圳纳米港有限公司 | 纯度≥95% |
| 高锰酸钾 | 上海化工院青磊特种气体有限公司 | 分析纯 |
| 硝酸钾 | 上海辰谊科学仪器有限公司 | 分析纯 |
| 氢氧化钾 | 江苏强盛化学股份有限公司 | 分析纯 |

注:制备过程中用到的仪器参见第 4 章表 4-2 实验仪器。

### 6.2.2 材料的制备

CGN 的制备包括氧化和还原两个过程。氧化过程是基于改性的 Hummers 法,2 g 碳纳米管和 1 g 硝酸钾在冰浴条件下溶解在 92 mL 浓硫酸中,搅拌 1 h 后,将 6 g 高锰酸钾缓慢加入其中,在室温下继续搅拌 1 h,加入 92 mL 水,搅拌 30 min 后,再加入 20 mL $H_2O_2$(30%)和 280 mL 水。洗涤、干燥后,取出 100 mg 所得样品分散在 100 mL 水中,随后,将 500 mg $NaBH_4$ 加入其中,搅拌 24 h,洗涤、干燥,即得到 CGN。

### 6.2.3 材料表征

1. 物理表征

实验中利用场发射扫描电子显微镜(FESEM；Philips XSEM30, Holland)和透射电镜(TEM；JEOL, JEM-2010,Japan)观测样品表面形貌,通过 X 射线衍射(XRD：D/Max-2400；Cu 靶；λ=1.541 8 Å；管电压 40 kV；管电流 60 mA；扫描速度 5°/min)和显微共聚焦拉曼光谱仪(514 nm laser with RM100)对样品进行了结构分析。氮气吸脱附曲线通过 Micromeritics Tristar 3 000 analyzer 在液氮温度下测试。

2. 电化学表征

将活性物质与乙炔黑和聚四氟乙烯按 15∶4∶1 的质量比混合制作电极[198]。先将前两者充分研磨后再加入聚四氟乙烯的乳液使其混合均匀后压在泡沫镍或石墨片集流体上作为测试电极。在室温下采用三电极测试系统在 CHI660d 电化学工作站上进行性能测试。所有测试以铂片做对电极,饱和甘汞做参比电极。

## 6.3 结果与讨论

通常,利用 Hummers 法可以将石墨有效地氧化成氧化石墨(GO),而氧化石墨也可以还原成还原氧化石墨烯(rGO)[199]。考虑到碳纳米管和石墨都是由石墨化的碳组成的,因此本章试图用类似的方法处理碳纳米管,

# 第6章 剪切多壁碳纳米管为弯曲石墨烯纳米片及其增强的电容性能

如图6-1所示。图6-2(a)和图6-2(c)为碳纳米管的扫描电镜图片,由图可知,其直径为40～100 nm,长度在微米级,相互缠绕在一起。经过氧化处

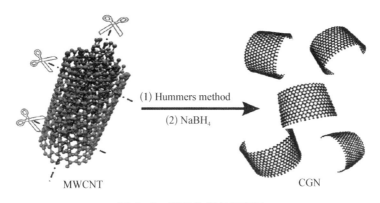

图6-1 CGN 的制备示意图

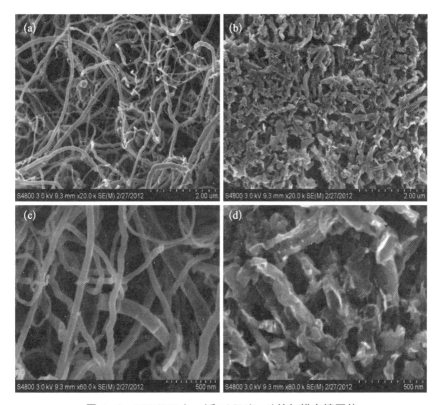

图6-2 MWCNTs(a,c)和 CGN(b,d)的扫描电镜图片

理后,形貌有了明显的改变,从低倍图片看到,碳纳米管被横向切断,长度约为几百纳米;从高倍图片观察到(图6-2(b)和图6-2(d)),这些纳米长度的碳纳米管被纵向打开,但仍保留了部分管状结构。透射电镜进一步证实了此种结构(图6-3);另外由高倍图片可以看出,与(002)晶面对应的层间距由于缺陷的存在而有所增加,从原始的0.34 nm至0.37 nm。

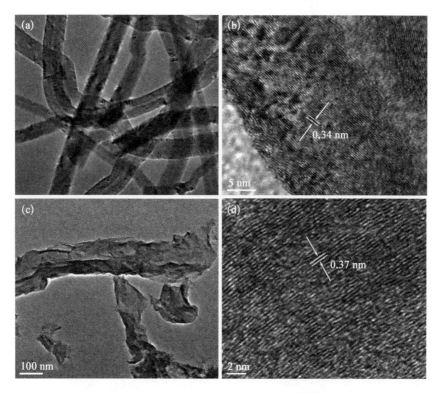

图6-3　MWCNTs(a,c)和CGN(b,d)的透射电镜图片

化学氧化导致碳纳米管表面出现了大量官能团,由 XRD 可以判断,如图6-4(a)所示,氧化后的碳纳米管的层间距由原来的3.4 Å变为8.6 Å。经过还原后,CGN 的层间距又变为3.7 Å,这与 TEM 结果一致。另外,较宽的峰也证明了 CGN 的无序结构。在 CGN 中,少量含氧官能团和无序度的出现有利于电解质离子的渗透。经过氧化后,碳纳米管表面出现一系列

红外吸收峰,如图6-4(b)所示,在3 400,1 624,1 725,1 065,1 245,1 375 cm$^{-1}$位置,分别对应于O—H伸缩,O—H弯曲,COOH的C=O,C—O,C—OH和C—O—C振动,这与氧化石墨的红外谱图一致[200]。拉曼光谱也进一步证实了氧化的过程,$I_D/I_G$值由原来的0.45变为1.88,表明强氧化造成了$sp^2$区域的减少。经过还原后,缺陷和残留官能团大幅减少,但仍有少量存在(图6-4(c))[201,202]。另外,比表面积有了明显的增加,碳纳米管和CGN的比表面积分别是47 m$^2$·g$^{-1}$和85 m$^2$·g$^{-1}$,此结果有力地证明了碳纳米管被成功剪切开。同时,也可以看到,碳纳米管的氧化与石墨的氧化非常类似[203,204]。

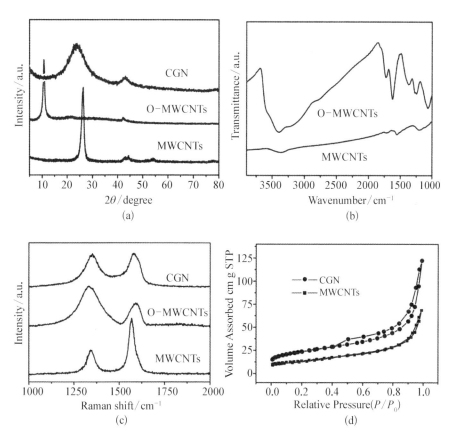

图6-4 样品的X射线衍射图谱(a)、红外图谱(b)、拉曼光谱(c)和氮气吸脱附曲线(d)

碳纳米管和 CGN 的电化学性能在三种电解质（1 M KOH，1 M $H_2SO_4$，1 M $Na_2SO_4$）中进行测试。从循环伏安曲线可以观察到（图 6-5），CGN 具有更大的面积，说明有更高的比电容。根据充放电曲线（图 6-6），

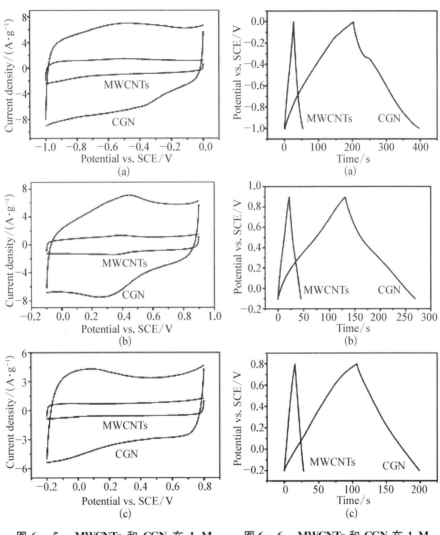

图 6-5　MWCNTs 和 CGN 在 1 M KOH(a)，1 M $H_2SO_4$(b)，1 M $Na_2SO_4$(c) 中的循环伏安曲线（50 mV·$s^{-1}$）

图 6-6　MWCNTs 和 CGN 在 1 M KOH(a)，1 M $H_2SO_4$(b)，1 M $Na_2SO_4$(c) 中的充放电曲线（1 A·$g^{-1}$）

# 第6章 剪切多壁碳纳米管为弯曲石墨烯纳米片及其增强的电容性能

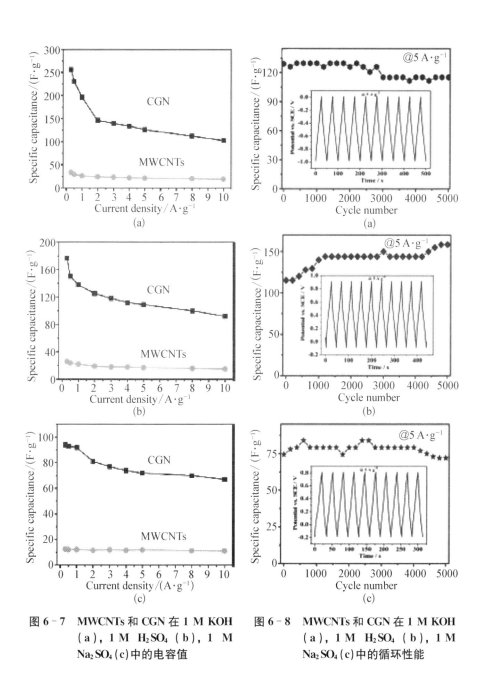

图 6-7 MWCNTs 和 CGN 在 1 M KOH (a)，1 M $H_2SO_4$ (b)，1 M $Na_2SO_4$ (c) 中的电容值

图 6-8 MWCNTs 和 CGN 在 1 M KOH (a)，1 M $H_2SO_4$ (b)，1 M $Na_2SO_4$ (c) 中的循环性能

计算出比电容值(图6-7)。在1 M KOH中电流密度为0.3 A·$g^{-1}$时,比电容值高达256 F·$g^{-1}$,远高于碳纳米管在相同条件下的电容值(33 F·$g^{-1}$)。在1 M $H_2SO_4$和1 M $Na_2SO_4$中,CGN具有更好的倍率性能。在电流密度为10 A·$g^{-1}$时,仍有71%电容保持率(相对于电流密度为0.3 A·$g^{-1}$)。CGN在不同电解质中表现出不同的电容值和倍率,可能与其表面的含氧官能团在不同电解质中的反应活性有关。CGN在三种电解质中的循环性能如图6-8所示,在1 M KOH中,经过5 000次循环后,电容仅衰减了1%;在1 M $H_2SO_4$中,电容值不但没有衰减反而有所增加;在1 M $Na_2SO_4$中,经过5 000次循环后,电容衰减了3%。这些结果表明,CGN具有优异的循环稳定性。

Andreas Hirsch[205,206]提出了碳纳米管氧化的机理。通常,碳的氧化有两种方法:Hummers法和Tour法。前者是在氧化过程中使用了硝酸钠而具有强氧化性;后者需要较高的温度(70℃),但是,反应物在加热的过程中(7 wt%/vol $KMnO_4$/$H_2SO_4$)容易爆炸。为了安全考虑,氧化最好在室温下进行。在我们的体系中,多壁碳纳米管在横向和纵向两个方向同时被切开。另外,从拉曼光谱和XRD图谱可以看到,CGN具有不完美或粗糙的边缘,这与以前文献报道一致[188,189]。表6-2总结了文献中有关碳纳米管超级电容器的电容性能。大多数碳纳米管的功率密度很高,但是其能量密度都很低。尽管单壁碳纳米管的电容值较高,但是昂贵的成本限制了其应用[207-210]。在表6-2中,不同的方法被用来增加多壁碳纳米管的能量密度。可是,能量的增加程度还是很有限[211-220]。本章所制备的CGN的比电容是目前文献报道有关碳纳米管比电容的最高值,主要归因于:① 碳纳米管的氧化和还原过程使其表面形成了少量含氧官能团会产生部分赝电容,一些结构缺陷有利于电解质离子渗透;② 横向切断使得电解质离子更容易传输;③ 纵向剪开释放了内在比表面积,从而提供了更多活性点;④ CGN具有一维碳纳米管和二维石墨烯片的独特性质。

表6-2 文献报道的关于碳纳米管的电容性能

| Technique | Nature of CNT | Electrolyte | Capacitance ($F \cdot g^{-1}$) | Ref (year) |
| --- | --- | --- | --- | --- |
| Spray deposition | Film/Multi, Single | 0.1 M $H_2SO_4$ | 77/155 | 207(2009) |
| CVD | Film/Multilayered | NaCl | 79.8 | 208(2010) |
| CVD | Film/Single | $EABF_4$/PC | 32 | 209(2009) |
| CVD | Film/Single | LiCl | 41.4 | 210(2010) |
| DC arc discharge | Powder/Single | 7.5 M KOH | 180 | 211(2001) |
| Electrostatic spray | Powder/Multi | 1 M $H_2SO_4$ | 108 | 212(2006) |
| CVD | Film/Multi | 6 M KOH | 20 | 213(2005) |
| KOH activation | Powder/Multi | 4 M $H_2SO_4$ | 62 | 214(2006) |
| $HNO_3$ oxidation | Powder/Double | 0.5 M $H_2SO_4$ | 54 | 215(2009) |
| CVD | Film/Single | 1 M $LiClO_4$/PC | 35 | 216(2011) |
| Layer-by-Layer Assembly | Film/Multi | 1 M $H_2SO_4$ | 159 | 217(2009) |
| Printable technique | Film/Single | $H_3PO_4$ | 120 | 218(2009) |
| Controlled oxidation | Powder/Single | $Et_4NBF_4$/PC | 114 | 219(2010) |
| Printable technique | Paper/Single | 1 M $LiPF_6$ | 200 | 220(2009) |
| The Tour method | Powder/Multi | 3 M NaOH | 165 | 196(2011) |
| The Hummers method | Powder/Multi | 1 M KOH | 256 | This work |

## 6.4 本章小结

利用Hummers法将多壁碳纳米管同时沿着横向—纵向剪切为卷曲的石墨烯纳米片。这种卷曲的石墨烯片具有一维纳米管和二维石墨烯的杂化结构。电化学测试表明,相比于多壁碳纳米管,卷曲的石墨烯片在酸性、碱性和中性电解质中均表现出更高的电容性能。例如，在电流密度为 $0.3\ \text{A}\cdot\text{g}^{-1}$ 时，比电容在碱性电解质中达到 $256\ \text{F}\cdot\text{g}^{-1}$。剪切后电容的增加，主要归因于较高的电解质润湿性、缺陷密度和比表面积。

# 第7章 三维的石墨烯/二氧化钒纳米带复合物凝胶的制备与电化学表征

## 7.1 引　言

近年来,随着能源和环境问题的日益突出,超级电容器因其优异的性质作为一种绿色储能装置填补了静电电容器和二次电池的空白[221-225]。根据电荷存储机理[226],超级电容器可以分为两类:双电层电容器和赝电容器。作为双电层电容器电极材料,碳材料具有电位窗口宽、功率密度高、循环稳定性能好和成本低等优点,但其比电容较低;过渡金属氧化物和导电聚合物作为赝电容器材料,通常具有高的比电容,但倍容率和稳定性较差且成本高是其致命的弱点[227-231]。通常,提高超级电容器电极材料性能的途径有:① 构建特定取向的纳米结构,具有比本体材料更优异的储能性质,例如纳米管[232-234]、纳米线[235,236]、纳米片[237,238]、纳米壁[239]等;② 设计复合材料,例如 $RuO_2@CNT$[240]、$Mn_3O_4@C$[241],以实现不同材料性能之间的协同效应。因此,构建一种多功能取向的纳米复合材料备受期待。

作为构建复合材料的一种理想的组分,石墨烯因具有优良的机械、电学性能、高比表面积和化学稳定性备受关注[242,243]。然而,石墨烯片层之间容易团聚而影响了固有的性质[244,245]。最近,三维的石墨烯水凝胶具有很好的应用[246]。石墨烯水凝胶具有多孔网络结构、多维的电子传导通道和离子润湿性[247-249]。通常,石墨烯凝胶具有中等的电容值(128～226 F·g$^{-1}$)[247-253]、优异的倍率性能和超长的稳定性。此外,氮或硼掺杂可以改善石墨烯的电容值[249,254,255]。为了进一步同时提高石墨烯凝胶的能量密度和功率密度,$Co_3O_4$纳米粒子被负载到石墨烯凝胶上[256],其比电容值在0～0.4 V电位窗口下,电流密度为0.5 A·g$^{-1}$时达到了757.8 F·g$^{-1}$。另外,graphene-Ni(OH)$_2$复合物凝胶在三电极体系下,比电容值在扫速为10 mV·s$^{-1}$时达1 250.3 F·g$^{-1}$[257]。尽管$Co_3O_4$和Ni(OH)$_2$的引入有效地增加了石墨烯凝胶的比电容值,但是钴镍材料的电位窗口都很小(0.4～0.5 V),进而限制了能量密度的进一步增加。

在各种过渡金属氧化物中,钒的氧化物因高容量、丰富氧化态和低成本而逐渐被人们关注。例如五氧化二钒的纳米带[258]、纳米纤维[259]、纳米管[260]、纳米线[261]等被应用于超级电容器的电极材料,其比电容值通常约为200 F·g$^{-1}$,且在水溶液中具有较高的电阻。当很薄的五氧化二钒薄膜沉积在碳纳米管上时[262,263],五氧化二钒才具有较高的比电容值。作为另外一种钒氧化物,二氧化钒具有比五氧化二钒更出色的电化学性能,这主要是源于钒的$V^{3+/5+}$混合价态和结构稳定性[264]。例如,$VO_2$/CMK-3已应用于超级电容器电极材料[265,266]。考虑到石墨烯凝胶具有比其他形式的碳更多的优势,例如高比表面积、电子和离子的通道、机械柔软性以及少量的含氧官能团等,因而非常期待二氧化钒纳米结构与石墨烯的复合物凝胶的电容性能。

在本章,通过一步水热法合成了三维多孔的石墨烯/二氧化钒纳米带复合物凝胶。在该凝胶结构中,二氧化钒纳米带与石墨烯纳米片相互交织,从而提供了更多的离子/电子通道。因此,该复合物凝胶表现出很高的电容性能。

## 7.2 实验部分

### 7.2.1 实验原料与仪器

实验过程中用到的原料和仪器如表 7-1 所示。

**表 7-1 实验原料**

| 原料名称 | 生产厂家 | 备注 |
| --- | --- | --- |
| $V_2O_5$ | Alfa Aesar | 分析纯 |
| SDS | 上海辰谊科学仪器有限公司 | 分析纯 |
| 硫酸钾 | 上海辰谊科学仪器有限公司 | 分析纯 |
| 无水乙醇 | 天津市福晨化学试剂厂 | 分析纯 |

注：制备过程中用到的仪器参见第 4 章表 4-2 实验仪器。

### 7.2.2 材料的制备

氧化石墨通过改性的 Hummers 方法制备而成[267,268]。120 mg 氧化石墨超声分散在 60 mL 水中，再加入 120 mg $V_2O_5$ 粉末，充分搅拌后，置于反应釜中，在 180℃加热 12 h，所得样品浸泡在二次水中 24 h 后，冷冻干燥。其中 $VO_2$ 的含量为 59%。在相同条件下也制备了石墨烯凝胶。二氧化钒纳米带制备过程如下[269]：将 $V_2O_5$(100 mg)和 SDS(500 mg)分散在 60 mL 水中，220℃加热 6 h，所得沉淀物洗涤干燥。

### 7.2.3 材料表征

**1. 物理表征**

实验中利用场发射扫描电子显微镜（FESEM；Philips XSEM30,

Holland)观测样品表面形貌,通过 X 射线衍射(XRD:D/Max - 2400;Cu 靶:λ=1.541 8 Å;管电压 40 kV;管电流 60 mA;扫描速度 5°/min)和显微共聚焦拉曼光谱仪(514 nm laser with RM100)对样品进行了结构分析;红外光谱测试通过仪器(Bruker Vertex 70 V)和衰减全反射(ATR)附件(Harrick Scientific Products,Ser no:GATVBR48406071201)得到。

2. 电化学表征

将上述得到的凝胶切成 1 mm 厚的薄片,作为工作电极,在室温下采用两电极测试系统在 CHI660d 电化学工作站上进行,所有测试均在 0.5 M $K_2SO_4$ 中进行。

## 7.3 结果与讨论

Graphene/$VO_2$ 纳米带复合物凝胶的形成过程如图 7-1 所示,氧化石墨烯作为基底原位生长 $VO_2$ 纳米带。在水热还原之前,氧化石墨烯纳米片由于具有很强的亲水性而分散在水中。当在 180℃水热还原的过程中,由

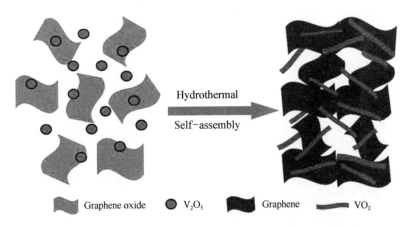

图 7-1 Graphene/$VO_2$ 纳米带复合物凝胶的形成过程

# 第7章 三维的石墨烯/二氧化钒纳米带复合物凝胶的制备与电化学表征

于含氧官能团的减少而逐渐变得疏水。憎水性、静电引力和π—π作用共同促使石墨烯片层之间相互交联[246]。在形成石墨烯凝胶的同时,五氧化二钒起初先溶解并分散在氧化石墨烯片上,最后逐渐生长且被氧化石墨烯还原成 $VO_2$ 纳米带结构,从而形成了复合物凝胶结构[270]。

图7-2是样品的XRD图,氧化石墨的层间距是8.6 Å,而石墨烯凝胶的层间距为3.7 Å,说明水热还原后大部分含氧官能团消失。$VO_2$ 纳米带和 Graphene/$VO_2$ 复合物凝胶的晶体结构一样,完全归属于 $VO_2$(B)(JCPDS No. 31-1438)[271]。在复合物中没有发现石墨烯凝胶的衍射峰,表明石墨烯片以单层形式均匀分散在复合物凝胶中。图7-3是样品的拉曼光谱图。将氧化石墨转化成石墨烯凝胶后,D峰与G峰的比例明显降低,也说明 $sp^2$ 区域增加。$VO_2$ 纳米带的拉曼信号与文献报道一致[272,273]。在复合物凝胶的拉曼光谱中,除了来自 $VO_2$ 的拉曼峰之外,石墨烯的特征峰(D峰和G峰)也随之出现,表明凝胶中两个组分同时存在。图7-4是石墨烯凝胶和 Graphene/$VO_2$ 复合物凝胶的衰减全反射红外光谱(ATR-FTIR)。对于石墨烯凝胶,在1 208 $cm^{-1}$ 和1 572 $cm^{-1}$ 位置的峰分别属于

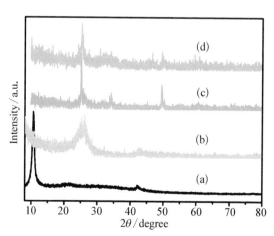

图7-2 (a) GO,(b) 石墨烯凝胶,(c) $VO_2$ 纳米带和
(d) Graphene/$VO_2$ 复合物凝胶的 XRD 图谱

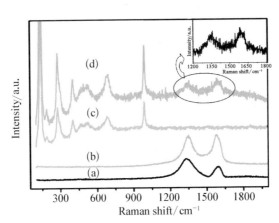

图7-3 (a) GO,(b) 石墨烯凝胶,(c) VO₂纳米带和(d) Graphene/VO₂复合物凝胶的拉曼光谱

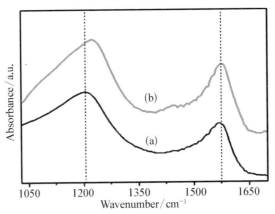

图7-4 (a) 石墨烯凝胶和(b) Graphene/VO₂复合物凝胶的衰减全反射红外光谱

C—OH 伸缩振动和石墨烯的骨架振动。在 Graphene/VO₂ 复合物凝胶中,C—OH 伸缩振动峰出现在 1 226 cm⁻¹ 处,相比于石墨烯凝胶,发生了明显的蓝移。这种现象间接证明了分子间氢键的存在。当 C—OH 的氢(H)与 VO₂ 中的强电负性的氧(O)发生键合时,来自石墨烯中的碳(C)与羟基中的氧(O)的作用力将变强,从而出现了 C—OH 伸缩振动峰的蓝移。这种氢键的形成有利于减少电极的极化现象,进而提高活性材料的利用率。

# 第7章 三维的石墨烯/二氧化钒纳米带复合物凝胶的制备与电化学表征

图 7-5 是 Graphene/VO$_2$ 复合物凝胶的 XPS 图谱。由图 7-5 可知，在测试范围内，C、V、O 等元素均被检测到。C1s(284.9 eV)对应于石墨烯中石墨化的碳。氧原子和钒原子的比例约为 2∶1，表明所制备的复合材料中钒氧化物为 VO$_2$。对于钒的 2p$_{3/2}$(517.1 eV)和 2p$_{1/2}$(524.4 eV)的结合能则与 VO$_2$ 中 V$^{4+}$ 相一致[274]。

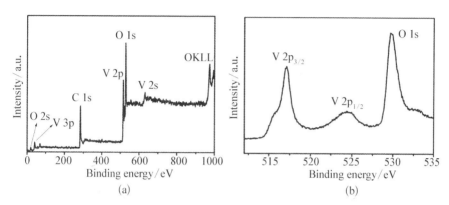

图 7-5　Graphene/VO$_2$ 复合物凝胶的 X 射线光电子能谱图

从图 7-6(a)可以看到，石墨烯凝胶具有三维多孔的网络结构，石墨烯片的超薄且柔性本质在高倍图片下可以清晰地观察到(图 7-6(b))。单纯的 VO$_2$ 的电镜图片如图 7-7 所示。可以看到 VO$_2$ 呈带状结构，其长度在微米级，宽度约为 60~150 nm。图 7-6(c)和图 7-6(d)为 Graphene/VO$_2$ 复合物凝胶在不同倍率下的扫描电镜图片：从低倍可以看到，复合物凝胶具有相互交织的微米级的孔结构；从高倍图片可以观察到，VO$_2$ 纳米带与石墨烯纳米片紧密连接在一起。VO$_2$ 与石墨烯之间的作用一方面减少了石墨烯的团聚问题，另一方面也增加了复合物凝胶的导电性。

图 7-8 是石墨烯凝胶和 Graphene/VO$_2$ 复合物凝胶的透射电镜图片。石墨烯凝胶中的石墨烯片在透镜下几乎呈透明状且有一些褶皱出现(图 7-8(a))。选区衍射进一步证明了石墨烯的高质量和单晶结构(图 7-8(b))。在复合物凝胶结构中，超薄的 VO$_2$ 纳米带直接生长在高度导电的石墨烯片上。这

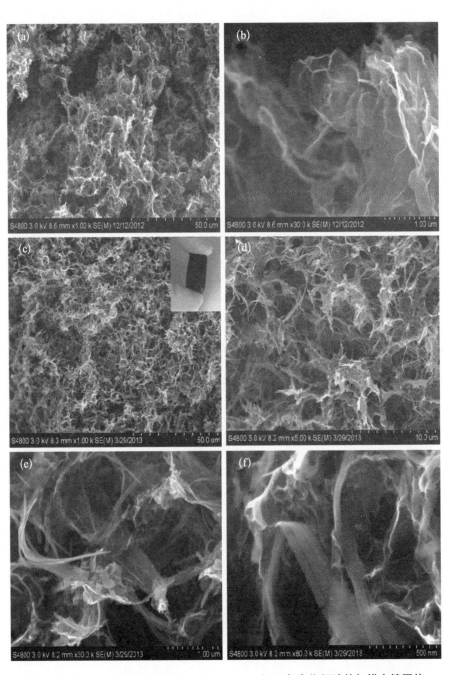

图7-6 (a,b) 石墨烯凝胶和(c-f) Graphene/VO$_2$复合物凝胶的扫描电镜图片

第 7 章　三维的石墨烯/二氧化钒纳米带复合物凝胶的制备与电化学表征

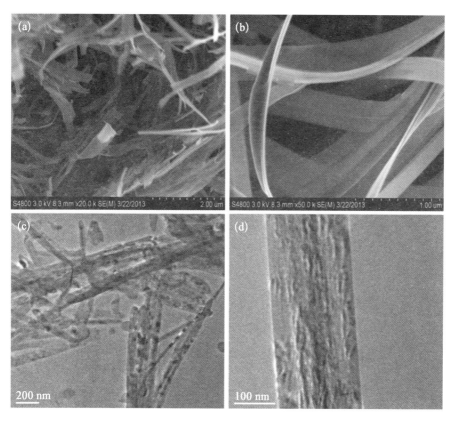

图 7-7　$VO_2$ 纳米带的扫描(a,b)和透射(c,d)电镜图片

些纳米带的宽度在 20～100 nm,长度约为数十微米。在高倍电镜下,间距为 0.58 nm 的晶格条纹属于 $VO_2(B)$ 的(002)晶面。此外,$VO_2$ 纳米带呈单晶结构,选区衍射中的衍射点完全对应于 $VO_2(B)$(JCPDS No. 31-1438)。

为了进一步考察在水热合成过程中氧化石墨烯与五氧化二钒的相互作用,我们在其他实验条件相同但没有氧化石墨的情况下水热处理五氧化二钒。如图 7-9 所示,处理前后,晶型没有任何的变化,而且从处理后的电镜图片来看(图 7-10),没有纳米带结构出现。这些结果间接地证明了在合成复合物凝胶过程中氧化石墨烯与五氧化二钒有强烈的相互作用。

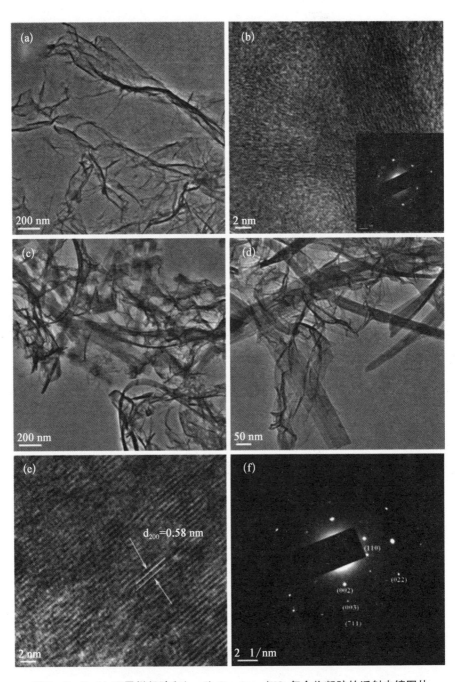

图 7-8 (a,b) 石墨烯凝胶和(c-f) Graphene/VO$_2$ 复合物凝胶的透射电镜图片

第 7 章 三维的石墨烯/二氧化钒纳米带复合物凝胶的制备与电化学表征

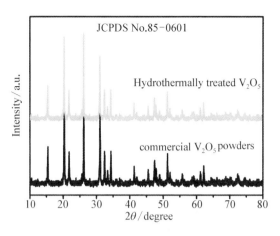

图 7-9 水热处理五氧化二钒前后的 X 射线衍射图

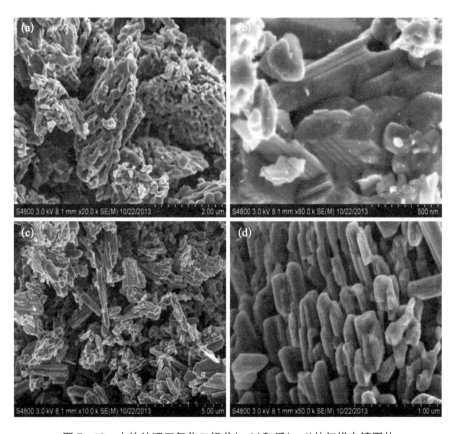

图 7-10 水热处理五氧化二钒前(a,b)和后(c,d)的扫描电镜图片

Graphene/VO$_2$ 复合物凝胶电化学性能在两电极体系中进行测试,相比于三电极体系,两电极体系更接近于实际应用[275]。在电解质的选择上,由于 K$^+$ 具有更小的水合离子半径和更高的离子电导率:K$^+$ =(0.33 nm,72.2 cm$^2$/$\Omega$ mol),Li$^+$ =(0.38 nm,38.6 cm$^2$/$\Omega$ mol),Na$^+$ =(0.358 nm,50.1 cm$^2$/$\Omega$ mol)[276],因此选用 K$_2$SO$_4$ 作为电解质。图 7-11 为石墨烯凝胶、VO$_2$ 纳米带和 Graphene/VO$_2$ 复合物凝胶的循环伏安曲线。石墨烯凝胶表现典型的双电层行为,VO$_2$ 纳米带和 Graphene/VO$_2$ 复合物凝胶的循环伏安曲线具有明显的氧化还原峰,相应的电极反应为

$$VO_2 + xK^+ + xe^- \longleftrightarrow K_xVO_2 \tag{1}$$

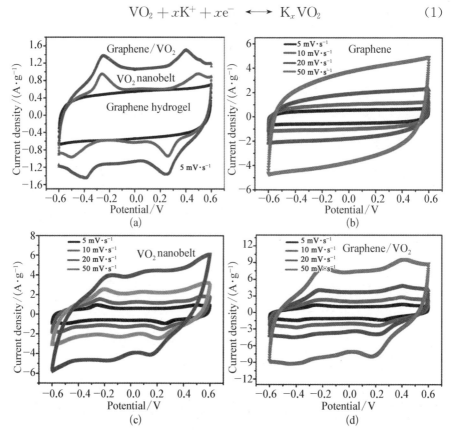

图 7-11 石墨烯凝胶、VO$_2$ 纳米带和 Graphene/VO$_2$ 复合物凝胶的循环伏安曲线

# 第7章 三维的石墨烯/二氧化钒纳米带复合物凝胶的制备与电化学表征

循环伏安曲线的积分面积的大小顺序为：Graphene/VO$_2$ 复合物凝胶＞VO$_2$ 纳米带＞石墨烯凝胶。面积越大，意味着比电容越高。

根据充放电曲线（图 7-12），按照计算公式

$$C = 2(I\Delta t / m\Delta V)$$

式中　$I$——放电电流；

　　　$\Delta t$——放电时间；

　　　$m$——单个电极的活性物质的质量；

　　　$\Delta V$——电位窗口。

石墨烯凝胶、VO$_2$ 纳米带和 Graphene/VO$_2$ 复合物凝胶在电流密度为 $1\ \text{A}\cdot\text{g}^{-1}$ 时的比电容值分别是 191，243，426 $\text{F}\cdot\text{g}^{-1}$。这与循环伏安曲线

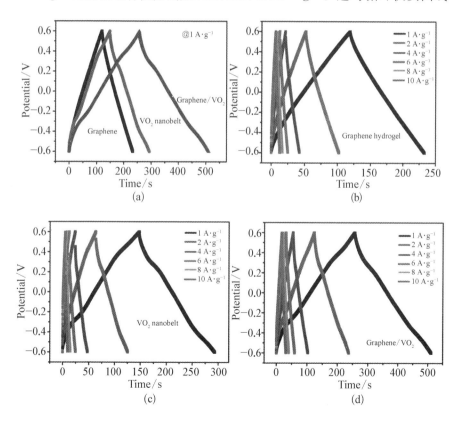

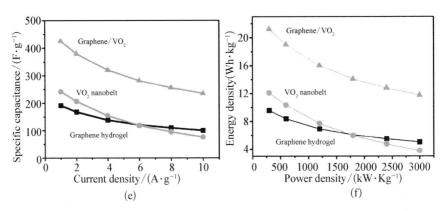

图 7-12 石墨烯凝胶、VO₂ 纳米带和 Graphene/VO₂ 复合物凝胶的充放电曲线和电容值

的测试结果一致。在电流密度为 10 A·g$^{-1}$ 时,复合物凝胶的电容值仍高达 239 F·g$^{-1}$,远高于 VO₂ 纳米带电流密度为 10 A·g$^{-1}$ 时的电容值。复合物凝胶的电容值甚至优于文献报道的 starfruit-like VO₂(218 F·g$^{-1}$)[265],VO₂/CMK-3 carbon(131 F·g$^{-1}$)[266] 和 hydrogenated VO₂(300 F·g$^{-1}$)[270]。

能量密度与功率密度的关系如图 7-12(f)所示,其计算公式分别是[21]:$E=[C(\Delta V)^2]/8$,$P=E/\Delta t$。复合物凝胶的最高能量密度可达到 21.3 Wh·kg$^{-1}$,远高于文献中的石墨烯凝胶的能量密度,例如 hydrazine reduced graphene hydrogels(5.7 Wh·kg$^{-1}$)[251],nitrogen-boron co-doped graphene aerogels(8.7 Wh·kg$^{-1}$)[249],和 functionalized graphene hydrogels(11.3 Wh·kg$^{-1}$)[277]。

为了进行更有效的对比,我们还测试了 Graphene VO₂ 混合物(其中,组分质量比与复合物凝胶相同)的电容性能,如图 7-13 所示。其循环伏安曲线的形状与复合物凝胶的类似。在电流密度为 1 A·g$^{-1}$ 时,比电容值是 227 F·g$^{-1}$。在电流密度为 10 A·g$^{-1}$ 时,仅有 80 F·g$^{-1}$,远低于复合物凝胶的电容值。这些结果进一步证明了在复合物凝胶中 VO₂ 纳米带与石墨烯之间的正协同效应。

样品的循环性能如图 7-14(a)所示。经过 5 000 次循环后,VO₂ 纳米

# 第 7 章　三维的石墨烯/二氧化钒纳米带复合物凝胶的制备与电化学表征

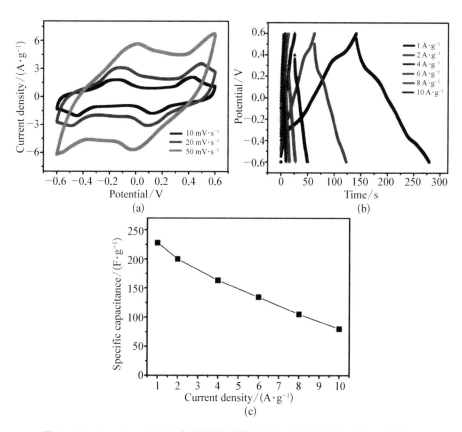

图 7-13　Graphene VO$_2$ 混合物的循环伏安(a)、充放电曲线(b)和电容值(c)

带损失了 41% 的电容值,这主要是由于钒氧化物的化学溶解和嵌入引起的结构塌陷[278]。石墨烯凝胶经过 5 000 次循环后仅衰减了 5%。复合物凝胶在起始的 1 000 圈内由于活化,电容值反而有所增加,经过 5 000 次循环后,电容值衰减了 8%。这个结果表明钒氧化物与石墨烯凝胶的复合可以有效地提高氧化钒的循环稳定性。通过交流阻抗进一步分析了电极反应的过程,如图 7-14(b)所示,半圆对应于电极的电荷传递电阻,随着石墨烯凝胶→Graphene/VO$_2$ 复合物凝胶→VO$_2$ 纳米带的次序,电荷传递电阻依次增大,这也进一步证明了当二氧化钒与石墨烯凝胶复合后,电极反应变得更快了。

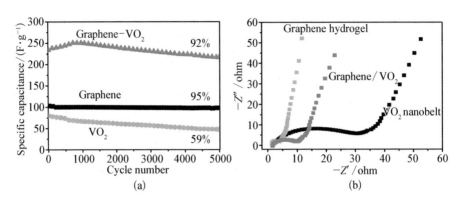

图 7‑14　石墨烯凝胶、$VO_2$ 纳米带、Graphene/$VO_2$ 复合物凝胶的循环性能(a)、交流阻抗(b)

## 7.4　本章小结

以五氧化二钒和氧化石墨为原料,通过一步法合成了三维石墨烯/二氧化钒纳米带复合物凝胶。在凝胶形成过程中,一维二氧化钒纳米带和二维石墨烯片通过氢键自组装成交联多孔的微结构。由于这种多孔凝胶结构和纳米带赝电容贡献,石墨烯/二氧化钒纳米带复合物凝胶在 $-0.6\sim 0.6$ V电位窗口下,比电容在电流密度为 $1\ A\cdot g^{-1}$ 时达到 $426\ F\cdot g^{-1}$,远大于相同测试条件下的单个组分的电容值($191\ F\cdot g^{-1}$ 和 $243\ F\cdot g^{-1}$)。此外,由于组分间的正协同效应,复合物凝胶电极表现出更高的倍率性能和循环稳定性。

# 第8章
# 单晶的三氧化二铁纳米粒子生长在石墨烯凝胶作为超级电容器负极材料

## 8.1 引 言

超级电容器因其高功率密度、快速充放电(几秒内)、良好的循环性能和较低的维修成本而在电化学能量储存方面具有独特的地位[279-281]。然而,现有的超级电容器的能量密度是非常有限的[282]。在不牺牲功率密度和循环寿命的前提下,增加超级电容器的能量密度是十分重要的。众所周知,电极材料是超级电容器的核心部分。在各种超级电容器电极材料中,赝电容材料(例如过渡金属氧化物和导电聚合物)具有比双电容材料(例如碳)更高的能量密度[282-285]。根据在水溶液中的电位窗口,赝电容材料包括两类:① 正极材料,电位窗口大于 0 V(vs. SCE);② 负极材料,电位窗口小于 0 V(vs. SCE)[286,287]。到目前为止,大多数文献报道主要集中于正极材料,而负极材料则很少有人关注。

通常,碳基材料应用于不对称电容器的负极材料[288,289]。但是,双电层限制了其比电容。因而寻求新型的负极材料将变得非常重要。一些金属氧化物,例如 $MoO_{3-x}$[290],$V_2O_5$[291],$TiN$[292],$VN$[293],$Bi_2O_3$[294] 和 $FeO_x$[295],被用来作为负极材料。其中,铁的氧化物由于具有丰富氧化态、低成本、无毒和环境友好而备受关注[296-310]。表8-1总结了氧化铁基超级电容器的电容性能。一般除了以薄膜形式存在时具有较好的性能外,铁氧化物因较差的导电性而限制了其在高电流密度下的比电容和倍率性能。为了进一步提高导电性和利用效率,文献报道了将铁氧化物与碳材料(例如碳纳米管、碳纤维、碳纳米片、碳泡沫、石墨烯等)进行复合。例如,$Fe_3O_4$纳米片/碳纤维的比电容达到135 $F \cdot g^{-1}$[303],远高于$Fe_3O_4$纳米片(83 $F \cdot g^{-1}$)。Wu等在文献中报道[306],当负载量为3 wt%时,$Fe_3O_4$的比电容值为510 $F \cdot g^{-1}$。最近,二维三明治结构的$Fe_3O_4$@graphene复合物表现出良好的倍率性能[310],但其比电容仍需进一步提高。因此,通过构建有效的离子/电子传输通道来提高铁氧化物的电容性能是很有必要的。

最近,三维的石墨烯水凝胶具有多孔网络结构、多维的电子传导通道和离子润湿性[246,247]。通常,石墨烯凝胶具有中等的电容值(128~226 $F \cdot g^{-1}$)、优异的倍率性能(100 $A \cdot g^{-1}$)和超长的稳定性[248-254]。为了进一步提高石墨烯水凝胶的电容性能,一些赝电容材料(例如$Co_3O_4$[256],$Ni(OH)_2$[321],$MnCO_3$[255]和$VO_2$[311])引入其中。可是,仍然没有铁氧化物/石墨烯凝胶作为超级电容器负极材料的文献报道。在本章,通过一步水热法制备了Graphene/$Fe_2O_3$复合物凝胶,将其应用于超级电容器负极材料时,具有超高的比电容值、优异的倍率性能和增强的循环性能。

## 8.2 实 验 部 分

### 8.2.1 实验原料与仪器

实验过程中用到的原料和仪器如表8-2所示。

# 第8章　单晶的三氧化二铁纳米粒子生长在石墨烯凝胶作为超级电容器负极材料

表8-1　文献中铁氧化物的电容性能总结

| $FeO_x$ electrode | Surface area ($m^2 \cdot g^{-1}$) | Electrolyte | Potential range (V vs. SCE) | Specific capacitance ($F \cdot g^{-1}$) | Rate capability | Ref (year) |
|---|---|---|---|---|---|---|
| $Fe_2O_3$ nanosheet film | Not reported | 1 M $Li_2SO_4$ | −0.9 to −0.1 V | 173 at 3 $A \cdot g^{-1}$ | 117 at 12.3 $A \cdot g^{-1}$ | 295(2009) |
| Cellular $Fe_3O_4$ film | Not reported | 1 M $Na_2SO_4$ | −0.55 to 0.05 V | 105 at 20 $mV \cdot s^{-1}$ | Not reported | 296(2005) |
| $Fe_3O_4$ powders | 115 | 0.1 M $K_2SO_4$ | −0.8 to 0.25 V | 75 at 10 $mV \cdot s^{-1}$ | Not reported | 297(2003) |
| $Fe_2O_3$ film | Not reported | 1 M NaOH | −0.6 to 0.1 V | 178 at 5 $mV \cdot s^{-1}$ | 121 at 100 $mV \cdot s^{-1}$ | 298(2011) |
| $Fe_2O_3$ nanotube arrays | Not reported | 1 M $Li_2SO_4$ | −0.8 to 0 V | 138 at 1.3 $A \cdot g^{-1}$ | 91 at 12.8 $A \cdot g^{-1}$ | 299(2011) |
| FeOOH rods | Not reported | 1 M $Li_2SO_4$ | −0.85 to −0.1 V | 116 at 0.5 $A \cdot g^{-1}$ | 93 at 1.5 $A \cdot g^{-1}$ | 300(2008) |
| Octadecahedron $Fe_3O_4$ film | Not reported | 1 M $Na_2SO_3$ | −1 to 0.1 V | 118 at 2 $A \cdot g^{-1}$ | 50 at 3.3 $A \cdot g^{-1}$ | 301(2009) |
| $Fe_3O_4$ nanoparticles | Not reported | 1 M $Na_2SO_3$ | −0.9 to 0.1 V | 207.7 at 0.4 $A \cdot g^{-1}$ | 90.4 at 10 $A \cdot g^{-1}$ | 302(2013) |
| $Fe_3O_4$/CNF composite | Not reported | 1 M $Na_2SO_3$ | −0.9 to 0.1 V | 127 at 10 $mV \cdot s^{-1}$ | 53 at 10 $mV \cdot s^{-1}$ | 303(2011) |
| $Fe_3O_4$/CNF composite | 229 | 6 M KOH | −1 to 0 V | 129 at 2.5 $mA \cdot cm^{-1}$ | 103 at 40 $mA \cdot cm^{-1}$ | 304(2013) |
| $Fe_3O_4$/carbon nanosheets | 34 | 1 M $Na_2SO_3$ | −0.8 to −0.2 V | 163.4 at 1 $A \cdot g^{-1}$ | 113 at 10 $A \cdot g^{-1}$ | 305(2013) |
| $Fe_3O_4$/carbon black | Not reported | 1 M $Na_2SO_4$ | −0.75 to 0.5 V | 510 for $Fe_3O_4$ at 15 $mA \cdot g^{-1}$ | Not reported | 306(2003) |
| $Fe_3O_4$ nanotubes/rGO | 431 | 2.5 M $Li_2SO_4$ | −1 to 0 V | 215 at 2.5 $mV \cdot s^{-1}$ | 88 at 100 $mV \cdot s^{-1}$ | 307(2012) |
| $FeO_x$ − carbon nanofoams | Not reported | 1 M KOH | −0.8 to 0.2 V | 343 for $FeO_x$ at 5 $mV \cdot s^{-1}$ | Not reported | 308(2010) |
| $Fe_3O_4$ particles-graphene | 160 | 1 M LiOH | −1 to 0.1 V | 220.1 at 0.5 $A \cdot g^{-1}$ | 134.6 at 5 $A \cdot g^{-1}$ | 309(2014) |
| FeOOH nanorods/graphene | | | −1.15 to 0.1 V | 326 at 0.5 $A \cdot g^{-1}$ | 293 at 10 $A \cdot g^{-1}$ | 310(2011) |

表 8-2 实 验 原 料

| 原 料 名 称 | 生 产 厂 家 | 备 注 |
| --- | --- | --- |
| $FeCl_3 \cdot 6H_2O$ | Alfa Aesar | 分析纯 |
| 氢氧化钾 | 江苏强盛化学股份有限公司 | 分析纯 |
| 无水乙醇 | 天津市福晨化学试剂厂 | 分析纯 |

注：制备过程中用到的仪器参见第 4 章表 4-2 实验仪器。

## 8.2.2 材料的制备

氧化石墨通过改性的 Hummers 方法制备而成[295,312]。120 mg 氧化石墨超声分散在 60 mL 水中，再加入 420 mg $FeCl_3 \cdot 6H_2O$ 粉末，充分搅拌后，置于反应釜中，180℃加热 12 h，所得样品浸泡在二次水中 24 h 后，冷冻干燥。其中 $Fe_2O_3$ 的含量为 65%。在相同条件下也制备了石墨烯凝胶和 $Fe_2O_3$。

## 8.2.3 材料表征

1. 物理表征

实验中利用场发射扫描电子显微镜（FESEM；Philips XSEM30，Holland）和透射电镜（TEM；JEOL，JEM-2010，Japan）观测样品表面结构。通过 X 射线衍射（XRD：D/Max-2 400；Cu 靶；λ＝1.541 8 Å；管电压 40 kV；管电流 60 mA；扫描速度 5°/min）和显微共聚焦拉曼光谱仪（514 nm laser with RM100）对样品进行了结构分析。氮气吸脱附曲线通过 Micromeritics Tristar 3 000 analyzer 在液氮温度下测试。红外光谱测试通过仪器（Bruker Vertex 70 V）和衰减全反射（ATR）附件（Harrick Scientific Products，Ser no：GATVBR48406071201）得到。

2. 电化学表征

将上述得到的凝胶切成 1 mm 厚的薄片，作为工作电极，在室温下采用三电极测试系统在 CHI660d 电化学工作站上进行。所有测试均在 1 M

KOH 中进行,以铂片做对电极,饱和甘汞做参比电极。

## 8.3 结果与讨论

Graphene/$Fe_2O_3$复合物凝胶的形成过程形成过程如图8-1所示,氧化石墨烯作为基底进行原位生长$Fe_2O_3$纳米粒子。在水热还原前,氧化石墨烯纳米片自由分散在水中。随后,$Fe^{3+}$通过静电作用与含氧官能团紧密连接。在水热过程中,$Fe^{3+}$首先形成FeOOH而沉积在氧化石墨烯纳米片上。随着水热的进行,氧化石墨烯被还原为石墨烯,同时FeOOH分解成$Fe_2O_3$。最终在氧化石墨烯与$Fe_2O_3$之间通过氢键作用,形成Graphene/$Fe_2O_3$复合物凝胶[246]。

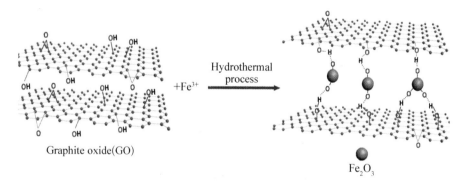

**图8-1 Graphene/$Fe_2O_3$复合物凝胶的形成过程**

通过XRD测试发现(图8-2),氧化石墨被还原,而形成的铁氧化物结构属于$Fe_2O_3$(JCPDS No. 33-0664)。石墨烯凝胶的衍射峰并没有出现在复合物凝胶的XRD上,说明石墨烯以单片形式均匀分散在三维复合物结构中。拉曼光谱如图8-3所示,将氧化石墨转化成石墨烯凝胶后,D峰与G峰的比例明显降低,也说明$sp^2$区域增加。$Fe_2O_3$纳米带的拉曼信号与文献报道一致[307]。在复合物凝胶的拉曼光谱中,除了来自$Fe_2O_3$的拉曼峰

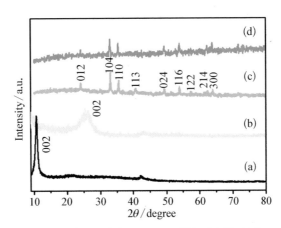

图 8-2 (a) GO,(b) Graphene 凝胶,(c) $Fe_2O_3$ 和
(d) Graphene/$Fe_2O_3$ 凝胶的 X 射线衍射图

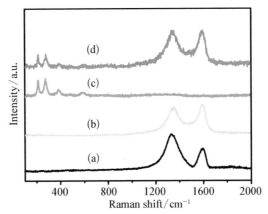

图 8-3 (a) GO,(b) Graphene 凝胶,(c) $Fe_2O_3$ 和
(d) Graphene/$Fe_2O_3$ 复合物凝胶的拉曼光谱

之外,石墨烯的特征峰(D 峰和 G 峰)也随之出现,表明凝胶中两个组分的同时存在。图 8-4 是石墨烯凝胶和 Graphene/$Fe_2O_3$ 复合物凝胶的衰减全反射红外光谱(ATR-FTIR)。对于石墨烯凝胶,在 1 208 cm$^{-1}$ 和 1 572 cm$^{-1}$ 位置的峰分别属于 C—OH 伸缩振动和石墨烯的骨架振动。在 Graphene/$Fe_2O_3$ 复合物凝胶中,C—OH 伸缩振动峰出现在 1 229 cm$^{-1}$ 处,相比于石墨烯凝胶,发生了明显的蓝移。这种现象间接证明了分子间氢键

# 第8章 单晶的三氧化二铁纳米粒子生长在石墨烯凝胶作为超级电容器负极材料

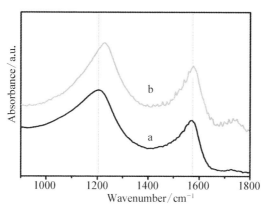

图 8-4 (a) Graphene 凝胶和(b) Graphene/$Fe_2O_3$ 复合物凝胶的 ATR-FTIR

的存在。当 C—OH 的氢(H)与 $Fe_2O_3$ 中的强电负性的氧(O)发生键合时，来自石墨烯中的碳(C)与羟基中的氧(O)的作用力将变强，从而出现了 C—OH 伸缩振动峰的蓝移。这种氢键的形成有利于减少电极的极化现象，进而提高活性材料的利用率[284]。

Graphene 凝胶、$Fe_2O_3$ 和 Graphene/$Fe_2O_3$ 复合物凝胶的氮气吸脱附曲线如图 8-5 所示，Graphene/$Fe_2O_3$ 复合物凝胶的比表面积高达 173 $m^2 \cdot g^{-1}$，远高于 Graphene 凝胶(134 $m^2 \cdot g^{-1}$)和 $Fe_2O_3$ (24 $m^2 \cdot g^{-1}$)。这个结果有力地证明了 $Fe_2O_3$ 纳米粒子修饰石墨烯片，可以适当阻止其团聚。另外，Graphene 凝胶、$Fe_2O_3$ 和 Graphene/$Fe_2O_3$ 复合物凝胶的孔容分别是 0.327 $cm^3 \cdot g^{-1}$、0.098 $cm^3 \cdot g^{-1}$、0.387 $cm^3 \cdot g^{-1}$。也就是，复合物凝胶的孔容远高于单纯的 $Fe_2O_3$ 的值，进一步表明 $Fe_2O_3$ 与石墨烯凝胶的复合有利于电解质离子的传输[313]。

如图 8-6(a)和图 8-6(b)所示，石墨烯凝胶具有三维多孔结构，且石墨烯片表现出透明的柔性特征。从图 8-6(c)和图 8-6(d)可以看到，在复合物凝胶中，$Fe_2O_3$ 纳米粒子(50～200 nm)均匀地修饰在石墨烯片上，形成了类似于石墨烯凝胶的多孔结构。在相同条件下(无氧化石墨存在)合成的三氧化二铁呈立方体特征，边缘尺寸为 50～100 nm。这种形貌的差异说

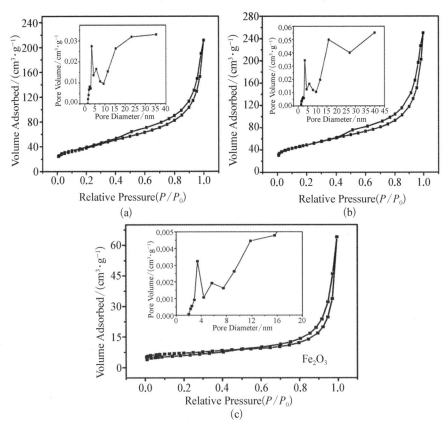

图 8-5 (a) Graphene 凝胶，(b) $Fe_2O_3$ 和 (c) Graphene/$Fe_2O_3$ 复合物凝胶的氮气吸脱附曲线

明在水热合成过程中，氧化石墨烯片与 $Fe^{3+}$ 之间的相互作用对于 $Fe_2O_3$ 的生长具有重要影响。

从透射电镜图片的对比，可以清晰地看到，$Fe_2O_3$ 纳米粒子均匀固定在石墨烯片上（图 8-6）。在高倍电镜下，间距为 0.37 nm 的晶格条纹属于 $Fe_2O_3$ 的(012)晶面。此外，$Fe_2O_3$ 呈单晶结构，选区衍射中的衍射点完全对应于 $Fe_2O_3$ (JCPDS No. 33-0664)。单纯的 $Fe_2O_3$ 在高倍电镜下，间距为 0.25 nm 的晶格条纹属于 $Fe_2O_3$ 的(110)晶面，也是单晶结构。不同的生长方向进一步解释了 $Fe_2O_3$ 在有无氧化石墨烯时的形貌差异。

# 第8章 单晶的三氧化二铁纳米粒子生长在石墨烯凝胶作为超级电容器负极材料

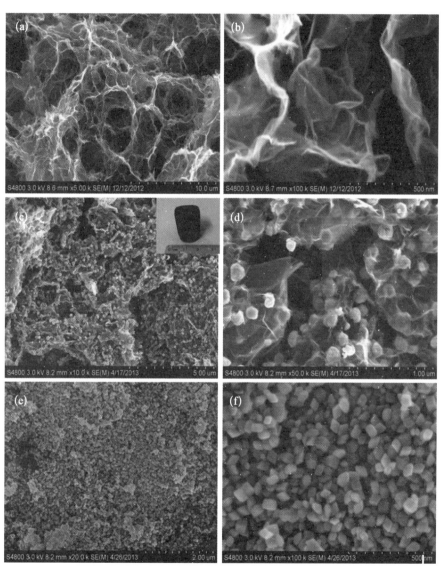

图 8-6 (a,b) Graphene 凝胶,(c,d) Graphene/Fe$_2$O$_3$ 复合物凝胶,(e,f) Fe$_2$O$_3$ 的扫描电镜图片

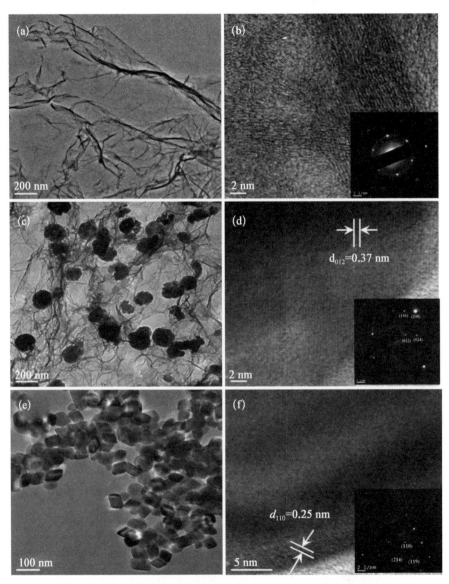

图 8-7 (a,b) Graphene 凝胶, (c,d) Graphene/$Fe_2O_3$ 复合物凝胶, (e,f) $Fe_2O_3$ 的 TEM 图片

# 第8章 单晶的三氧化二铁纳米粒子生长在石墨烯凝胶作为超级电容器负极材料

图 8-8 是 Graphene 凝胶、$Fe_2O_3$ 和 Graphene/$Fe_2O_3$ 复合物凝胶在 $(-1.2\sim-0.3\text{ V})$ 电位窗口下的循环伏安曲线。当低于 $-1\text{ V}$ 时，石墨烯凝胶的析氢现象变得非常明显。在 $(-1\sim-0.3\text{ V})$ 电位窗口下，石墨烯凝胶的循环伏安曲线呈现矩形特征，属于典型的双电层电容。Graphene/$Fe_2O_3$ 复合物凝胶的循环伏安曲线有一对明显的氧化还原峰，对应于 $Fe^{2+}$ 和 $Fe^{3+}$ 之间的转化。当负向扫描时，$Fe^{3+}$ 在析氢之前被还原为 $Fe^{2+}$；当正向扫描时，$Fe^{2+}$ 被氧化为 $Fe^{3+}$[314]。在高的扫描速率下，其形状仍保持不变，说明复合物凝胶具有良好的电容性能。$Fe_2O_3$ 的循环伏安曲线与复合物凝胶的非常相似，但其循环伏安曲线的积分面积明显小于复合物凝胶，

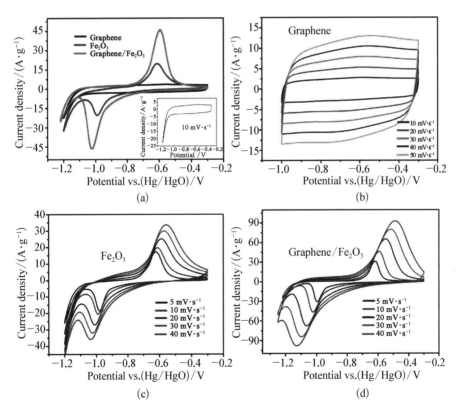

图 8-8 样品的循环伏安曲线

表明 Graphene/$Fe_2O_3$ 复合物凝胶具有更高的比电容[315]。

Graphene 凝胶、$Fe_2O_3$ 和 Graphene/$Fe_2O_3$ 复合物凝胶的充放电曲线如图 8-9 所示,为了避开析氢反应的影响,电位窗口选为 $-1.05\sim-0.3$ V。石墨烯凝胶的电容值在电流密度为 2 A·$g^{-1}$ 时达 272 F·$g^{-1}$,在电流密度为 50 A·$g^{-1}$ 时电容保持率为 55%。对于 Graphene/$Fe_2O_3$ 复合物凝胶,电容值在电流密度为 2 A·$g^{-1}$ 时高达 908 F·$g^{-1}$,在 50 A·$g^{-1}$ 时电容仍有 622 F·$g^{-1}$,这是到目前为止对于铁氧化物基超级电容器研究最好的结果。对比之下,$Fe_2O_3$ 在电流密度为 20 A·$g^{-1}$ 时电容值仅为 91 F·$g^{-1}$,约为 2 A·$g^{-1}$ 时的 10%。复合物凝胶的高电容性能归因于两组分之间的协同效应。鉴于 Graphene/$Fe_2O_3$ 复合物凝胶的电位窗口和高比电容值,再选择

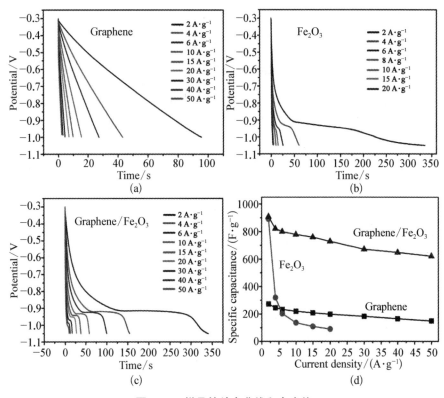

图 8-9 样品的放电曲线和电容值

第 8 章　单晶的三氧化二铁纳米粒子生长在石墨烯凝胶作为超级电容器负极材料

适当的正极材料（例如 $MnO_2$[316]，$Ni(OH)_2$[317]，$NiCo_2O_4$[279]），有望可以同时实现超高的能量密度和功率密度。

$Fe_2O_3$ 和 Graphene/$Fe_2O_3$ 复合物凝胶的循环性能在 $20\ mV \cdot s^{-1}$ 下进行，结果如图 8-10 所示。从循环过程中前 5 圈的曲线可以明显看到，$Fe_2O_3$ 的衰减程度远大于 Graphene/$Fe_2O_3$ 复合物凝胶。对于 $Fe_2O_3$ 样品，经过 70 次循环后，电容损失了 51%；而对于复合物凝胶，经过 200 次循环后，电容仍能保持 75%。循环性能的明显提高主要是由于：当 $Fe_2O_3$ 纳米粒子沉积在三维多孔的石墨烯凝胶上时，可以减少铁在循环过程中的溶解，同时石墨烯凝胶可以作为基质保持 $Fe_2O_3$ 的结构完整。

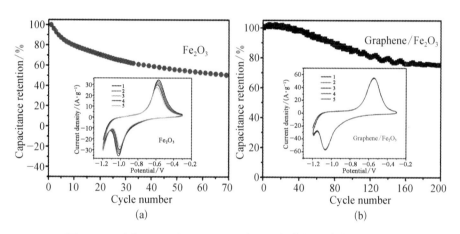

图 8-10　(a) $Fe_2O_3$、(b) Graphene/$Fe_2O_3$ 复合物凝胶的循环性能

## 8.4　本章小结

单晶的三氧化二铁纳米粒子直接生长在石墨烯凝胶上，作为高性能的超级电容器负极材料。在三氧化二铁/石墨烯复合物凝胶形成过程中，三氧化二铁纳米粒子与石墨烯片通过氢键作用自组装形成高表面积多孔结

构。在$-1.05\sim-0.3$ V电位窗口下,复合物凝胶具有高比电容(电流密度为$2\ A\cdot g^{-1}$时达$908\ F\cdot g^{-1}$)和优异的倍率性能(电流密度为$100\ A\cdot g^{-1}$时,电容保持率为68%)。另外,相比于单纯的三氧化二铁,复合物凝胶的稳定性也有明显的提高。

# 第9章

# 基于石墨烯凝胶正极和二氧化钛纳米带阵列负极的杂化超级电容器的构建及其超高的能量密度

## 9.1 引　　言

近年来,随着能源枯竭和全球变暖,能量储存将变得十分重要[318-320]。在现有的能量储存装置中,锂离子电池和超级电容器是最有前景的[321]。这两种装置各有优缺点且优势互补。前者具有很高的能量密度,而后者具有较高的功率密度[322-329]。如果构建一种储能装置综合二者的优势,将是十分有意义的。

最近,杂化电容器的出现实现了上述愿望,在有机电解质中一个电极用超级电容器电极材料,而另一个电极用锂离子电池材料[330-334]。通常,钛基化合物用作杂化电容器负极材料,尽管有一系列进展,但其功率密度因两电极之间的不平衡而变得很低[332-339]。通常,阵列电极具有较好的离子/电子传输过程[320-342],但是关于纳米阵列在杂化电容器中的应用很少报道。对于杂化电容器的正极而言,最常用的材料是活性炭,但其倍率性能十分有限[333,343,356],因此,设计合成一种具有三维多孔网络结构的碳材料对于提高杂化电容器的性能也是十分必要的。三维的石墨烯水凝胶具有多孔网络结构、多维的电子传导通道和离子润湿性[246]。若将该水凝胶作为杂化电容器的正极对于提高其储能性质是十分期待的。

在本章,首先制备了 $TiO_2$ 纳米带阵列;其次,合成了三维石墨烯水凝胶;最后,以 $TiO_2$ 纳米带阵列为正极,石墨烯水凝胶为负极,在有机电解质中构建了一种新型的杂化电容器,其能量密度高达 82 Wh·$kg^{-1}$。

## 9.2 实验部分

### 9.2.1 实验原料与仪器

实验过程中用到的原料和仪器如表 9-1 所示。

表 9-1 实验原料

| 原料名称 | 生产厂家 | 备注 |
| --- | --- | --- |
| 钛箔 | Alfa Aesar | 分析纯 |
| 浓盐酸 | 江苏强盛化学股份有限公司 | 分析纯 |
| 氢氧化钠 | 江苏强盛化学股份有限公司 | 分析纯 |

注:制备过程中用到的仪器参见第 4 章表 4-2 实验仪器。

### 9.2.2 材料的制备

$TiO_2$ 纳米带阵列的制备过程如下:将钛箔(2.0 cm×4.0 cm)浸泡在 NaOH 中,置于水热反应釜中,封盖,160℃加热 18 h,等冷却至室温后,取出钛箔,洗涤、干燥后,浸入 1 M HCl 中 1 h,最后在空气气氛下 500℃煅烧 2 h。石墨烯凝胶的制备参考文献[345]～文献[347]。

## 9.2.3 材料表征

**1. 物理表征**

实验中利用场发射扫描电子显微镜(FESEM；Philips XSEM30，Holland)观测样品表面形貌，通过 X 射线衍射(XRD：D/Max－2400；Cu 靶：$\lambda=1.5418$ Å；管电压 40 kV；管电流 60 mA；扫描速度 5°/min)和显微共聚焦拉曼光谱仪(514 nm laser with RM100)对样品进行了结构分析。

**2. 电化学表征**

首先在半电池中测试单个电极的性能，电解质是 1 M $LiPF_6$/EC－DMC，锂箔作为对电极和参比电极。杂化电容器的构建如下：以 $TiO_2$ 纳米带阵列为负极，石墨烯水凝胶为正极，利用聚丙烯膜作为隔膜，电解质同上，其中，正负极的质量比为 3∶1。

## 9.3 结果与讨论

在水热还原之前，氧化石墨的层间距是 8.6 Å，而还原后的层间距为 3.7 Å，说明大部分含氧官能团消失，成为而石墨烯凝胶(图 9－1(a))。将氧化石墨转化成石墨烯凝胶后，D峰与G峰的比例明显降低，也说明 $sp^2$ 区域增加，意味着导电性能的增加(图 9－1(b))。从扫描图片可以看到，石墨烯凝胶呈三维多孔网络结构。透射电镜和选区衍射进一步证明了石墨烯凝胶的高质量(图 9－2)[348]。电化学性能如图 9－3 所示，循环伏安曲线呈矩形，充放电曲线呈线性关系，属于典型的双电层超级电容器行为，在电流密度为 0.5 A·$g^{-1}$ 时，电容(或容量)为 133 F·$g^{-1}$(52 mAh·$g^{-1}$)，远高于活性炭在该区间的容量[349,350]。在电流密度为 10 A·$g^{-1}$ 时，容量保持率仍有 75%，且经过 1 000 次循环后，容量损失了 4%。这些良好的电容性能可

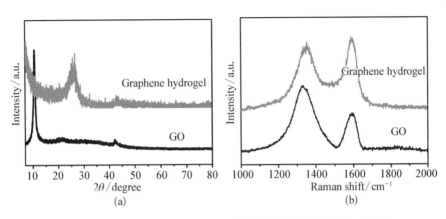

图 9-1 GO 和 Graphene 凝胶的 X 射线衍射图片(a)和拉曼光谱(b)

图 9-2 Graphene 凝胶的扫描电镜图片(a,b),照片(c)和透射电镜图片(d)

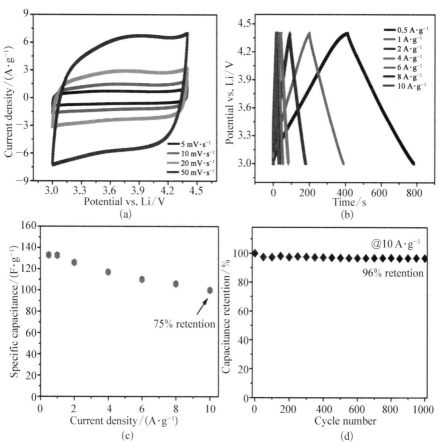

图9-3 石墨烯凝胶在(3.0~4.4 V vs. Li)的循环伏安曲线(a),
充放电曲线(b),电容值(c)和循环性能(d)

以为杂化电容器提供较高的功率密度。

$TiO_2$纳米带阵列的XRD如图9-4所示,除了基底钛的衍射峰之外,其余的衍射峰全部归属于$TiO_2$(JCPDS No. 21-1272)。从扫描电镜图片可以看出(图9-5),$TiO_2$纳米带均匀垂直生长在钛基底上,其宽度约为400 nm,长度约为8 μm,厚度约为20 nm。高倍TEM下,0.35 nm晶格间距对应于(101)晶面(图9-5(e))。选区衍射证明了单晶结构和晶体结构$TiO_2$(JCPDS No. 21-1272)。在电流密度0.1 A·$g^{-1}$时,其容量为160 mAh·$g^{-1}$(接近于理论容量)[351,352]，在电流密度高达 5 A·g$^{-1}$时，其容量仍有 43 mAh·g$^{-1}$。在电流密度 0.2 A·g$^{-1}$时，循环 100 圈后，容量保持 85%(图 9-6)。这些优异的性能可以为杂化电容器提供较高的能量[353-356]。

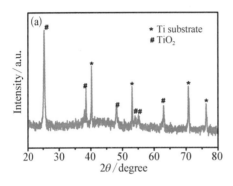
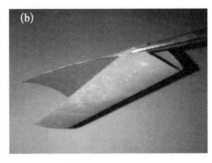

图 9-4　TiO$_2$纳米带阵列的 X 射线衍射图(a)和照片(b)

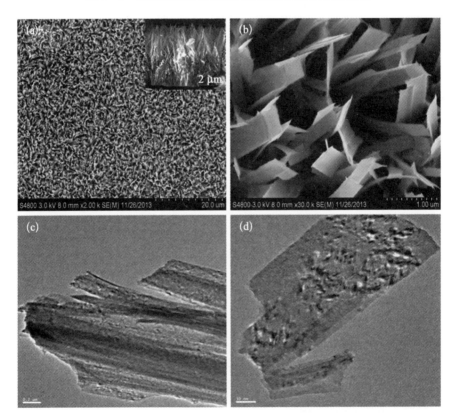

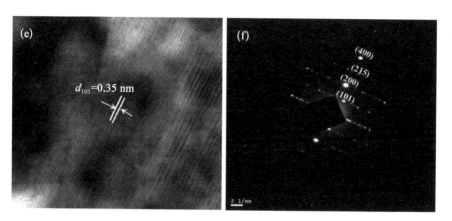

图9-5 TiO₂纳米带阵列的扫描电镜图片(a,b),透射电镜图片(c-e),选区衍射(f)

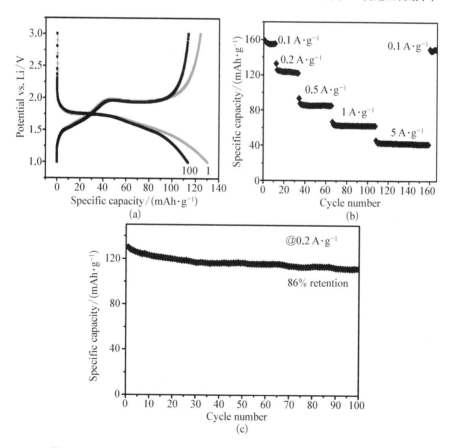

图9-6 TiO₂纳米带阵列的充放电曲线(a),容量值(b)和循环性能(c)

杂化电容器的构建如图 9-7 所示,在 1 M LiPF$_6$/EC-DMC 电解质中,以石墨烯凝胶为正极、TiO$_2$ 纳米带阵列为负极。图 9-8(a)是其循环伏安曲线,由于负极 TiO$_2$ 的氧化还原活性,循环伏安曲线具有明显的峰,电位窗口可以达到 3.8 V。循环性能介于锂离子电池和超级电容器之间,经过 600 次循环后,容量保持率为 73%。根据充放电曲线,按照公式:$P = \Delta E i/m$,$E = Pt$ 和 $\Delta E = (E_{max} + E_{min})/2$,可以计算出能量密度和功率密度。最高能量密度高达 82 Wh·kg$^{-1}$,最高功率密度高达 19 kW·kg$^{-1}$(一个完整的充放电仅需 8.4 s)。如此高的能量密度和功率密度远超过一些文献报道的值,例如 LiTi$_2$(PO$_4$)$_3$//AC (14 Wh·kg$^{-1}$—0.18 kW·kg$^{-1}$)[322],TiO$_2$-B Nanowire//CNT (12.5 Wh·kg$^{-1}$—10 C)[335],V$_2$O$_5$ fibers//Bucky paper (18 Wh·kg$^{-1}$ 和 315 kW·kg$^{-1}$)[357],TiO$_2$-B nanorods//AC (23 Wh·kg$^{-1}$ 和 2.8 kW·kg$^{-1}$)[350],TiO$_2$-B nanotubes//MWCNTs (19.3 Wh·kg$^{-1}$—10 C)[358],Li$_4$Ti$_5$O$_{12}$//AC system (1 000~2 000 W·kg$^{-1}$—10~15 Wh·kg$^{-1}$)[332],F-Fe$_2$O$_3$/AC (28 Wh·kg$^{-1}$—0.55 kW·kg$^{-1}$)[359],TiP$_2$O$_7$//AC (13 Wh·kg$^{-1}$ 和 371 W·kg$^{-1}$)[360],TiO$_2$-(还原氧化石墨烯)//AC (8.9 Wh·kg$^{-1}$—8 kW·kg$^{-1}$)[361] 和 CNT/V$_2$O$_5$//AC

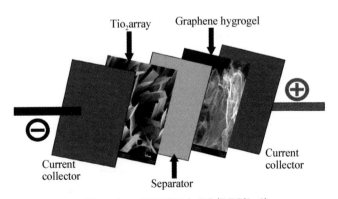

图 9-7 杂化电容器示意图

# 第9章 基于石墨烯凝胶正极和二氧化钛纳米带阵列负极的杂化超级电容器的构建及其超高的能量密度

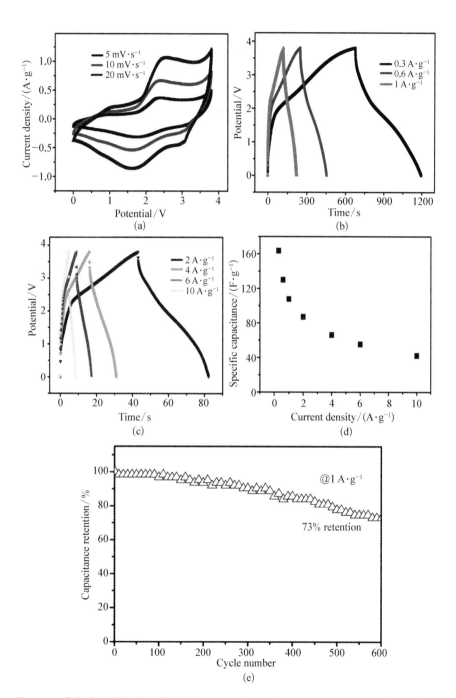

图9-8 杂化电容器的循环伏安曲线(a),充放电曲线(b,c),电容值(d)和循环性能(e)

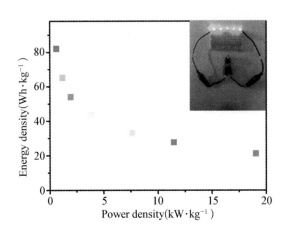

图 9-9　杂化电容器的能量-功率关系图

$(6.9\ Wh \cdot kg^{-1}$—$6.3\ kW \cdot kg^{-1})$[373]。同时,本章所制备的杂化电容器能够同时点亮 5 个 LED 灯,说明具有潜在的应用价值。

## 9.4　本章小结

本章以石墨烯凝胶为正极,二氧化钛纳米带为负极,$LiPF_6/EC-DMC$ 为有机电解质构建了一种新型的杂化电容器。由于石墨烯凝胶的多孔性、高导电性和二氧化钛独特的纳米带阵列结构,杂化电容器有利于快速的离子和电子的传输。在 0~3.8 V 电位窗口下,能量密度高达 82 $Wh \cdot kg^{-1}$。甚至在 8.4 s 充放电倍率下,能量密度仍保持有 21 $Wh \cdot kg^{-1}$。这些测试结果表明,该杂化电容器具有比超级电容器更高的能量密度和比锂离子电池更高的功率密度。

# 第10章 结论与展望

## 10.1 结 论

本书采用脉冲激光沉积、水热、化学气相沉积和电化学沉积等方法设计合成了一系列具有新型纳米结构的超级电容器电极材料,同时对材料的结构、形貌、形成机理进行了研究。实验所得不同结构材料有:多孔氧化镍薄膜、二氧化锰纳米片阵列、氧化镍/石墨烯泡沫、多孔氮掺杂的碳纳米管、钴酸镍纳米线和纳米片/碳布、卷曲的石墨烯纳米片、石墨烯/二氧化钒纳米带复合物凝胶、三氧化二铁/石墨烯复合物凝胶、石墨烯凝胶、二氧化钛纳米带阵列。这些材料表现出优异的超级电容性能,特别是具有出色的倍率性能(或功率密度)。通过构建不对称超级电容器和杂化超级电容器,显著地提高了超级电容器的能量密度。

## 10.2 展 望

本书的研究工作是在超级电容器电极材料的设计合成及其电容性能

研究方面作了一个初步的探索，在实际应用中仍有许多理论及技术问题尚待解决，相信随着电极材料研究与制造技术的不断深入和发展，上述电极材料在超级电容器中的应用将会取得突破。在以后的工作中，在以下几个方面还有待进一步考察：

（1）本书中采用的脉冲激光沉积薄膜制备技术也适用于制备其他单一金属氧化物薄膜、金属氮化物薄膜、金属碳化物薄膜、多层薄膜结构、异质金属氧化物核壳薄膜结构和石墨烯薄膜结构等，可以进行更多和更深入的研究探讨；

（2）通常，脉冲激光沉积制备的薄膜具有致密的结构，如何结合其他的实验手段制备出多孔疏松的纳米结构也是一种很有趣的尝试；

（3）在材料表征部分，几乎所有的测试（例如 X 射线衍射、扫描/透射电镜、比表面积测试等）都是在材料合成之后进行的，如果能够对材料表面的电极反应机理进行原位观察和分析，将是以后工作的亮点；

（4）对材料的设计以及相应的电极反应机理进行理论模拟计算，揭示结构与性能、组成与性能、组分间相互作用力与性能以及尺寸和形貌与性能之间的关系，将是储能领域的研究趋势。

# 参考文献

[1] Stoller M D, Ruoff R S. Best practice methods for determining an electrode material's performance for ultracapacitors[J]. Energy Environ. Sci., 2010, 3(9): 1294-1301.

[2] Winter M, Brodd R J. What are batteries, fuel cells, and supercapacitors? [J]. Chem. Rev. 2004, 104(10): 4245-4269.

[3] Zhang L L, Zhao X S. Carbon-based materials as supercapacitor electrodes[J]. Chem. Soc. Rev., 2009, 38(9): 2520-2531.

[4] Yu G, Xie X, Pan L, et al. Hybrid nanostructured materials for high-performance electrochemical capacitors [J]. Nano Energy, 2013, 2(2): 213-234.

[5] Zhai Y, Dou Y, Zhao D, et al. Carbon materials for chemical capacitive energy storage[J]. Adv. Mater., 2011, 23(42): 4828-4850.

[6] Gupta V, Gupta S, Miura N. Potentiostatically deposited nanostructured $Co_xNi_{1-x}$ layered double hydroxides as electrode materials for redox-supercapacitors[J]. J. Power Sources, 2008, 175(1): 680-685.

[7] Hu Z A, Xie Y L, Wang Y X, et al. Synthesis of α-cobalt hydroxides with different intercalated anions and effects of intercalated anions on their morphology, basal plane spacing, capacitive property[J]. J. Phys. Chem. C,

2009，113(28)：12502－12508.

[8] Hu Z A, Xie Y L, Wang Y X, et al. Synthesis and electrochemical characterization of mesoporous $Co_x Ni_{1-x}$ layered double hydroxides as electrode materials for supercapacitors[J]. Electrochim. Acta, 2009 (10)：54, 2737－2741.

[9] Choi D, Blomgren G E, Kumta P N. Fast and reversible surface redox reaction in nanocrystalline vanadium nitride supercapacitors[J]. Adv. Mater., 2006, 18(9)：1178－1182.

[10] Qu Q, Yang S, Feng X. 2D sandwich-like sheets of iron oxide grown on graphene as high energy anode material for supercapacitors[J]. Adv. Mater., 2011, 23(9)：5574－5580.

[11] Zhu Y, Murali S, Stoller M D, et al. Carbon-based supercapacitors produced by activation of graphene[J]. Science, 2011, 332(6037)：1537－1541.

[12] Wang D W, Li F, Liu M, et al. 3D Aperiodic hierarchical porous graphitic carbon material for high-rate electrochemical capacitive energy storage[J]. Angew. Chem. Int. Ed., 2008, 47(2)：373－376.

[13] Ning G, Fan Z, Wang G, et al. Gram-scale synthesis of nanomesh graphene with high surface area and its application in supercapacitor electrodesw[J]. Chem. Commun., 2011, 47(21)：5976－5978.

[14] Chmiola J, Yushin G, Gogotsi Y, et al. Anomalous increase in carbon capacitance at pore sizes less than 1 nanometer[J]. Science, 2006, 313(5794)：1760－1763.

[15] Worsley M A, Olson T Y, Lee J R I, et al. High surface area, $sp^2$-cross-linked three-dimensional graphene monoliths[J]. J. Phys. Chem. Lett., 2011, 2(8)：921－925.

[16] Li Y, Li Z, Shen P K. Simultaneous formation of ultrahigh surface area and three-dimensional hierarchical porous graphene-like networks for fast and highly stable supercapacitors[J]. Adv. Mater., 2013, 25(17)：2574－2480.

[17] Xia X, Zhang Y Q, Chao D, et al. Solution Synthesis of Metal Oxides for Electrochemical Energy Storage Applications[J]. Nanoscale, 2014, 6: 5008-5048.

[18] Ren Y, Ma Z, Bruce P G. Ordered mesoporous metal oxides: synthesis and applications[J]. Chem. Soc. Rev., 2012, 41(14): 4909-4927.

[19] Hu C C, Chang K H, Lin M C, et al. Design and tailoring of the nanotubular arrayed architecture of hydrous $RuO_2$ for next generation supercapacitors[J]. Nano Lett., 2006, 6(12): 2690-2695.

[20] Lang X Y, Hirata A, Fujita T, et al. Nanoporous metal/oxide hybrid electrodes for electrochemical supercapacitors[J]. Nat. Nanotech., 2011, 6(4): 232-236.

[21] Yang L, Cheng S, Ding Y, et al. Hierarchical network architectures of carbon fiber paper supported cobalt oxide nanonet for high-capacity pseudocapacitors[J]. Nano Lett., 2012, 12(1): 321-325.

[22] Ghosh A, Ra E J, Jin M, et al. High pseudocapacitance from ultrathin $V_2O_5$ films electrodeposited on self-standing carbon-nanofiber paper[J]. Adv. Funct. Mater., 2011, 21(13): 2541-2547.

[23] Wu H Y, Wang H W. Electrochemical synthesis of nickel oxide nanoparticulate films on nickel foils for high-performance electrode materials of supercapacitors [J]. Int. J. Electrochem. Sci., 2012, 7: 4405-4417.

[24] Cho S I, Lee S B. Fast electrochemistry of conductive polymer nanotubes: synthesis, mechanism and application[J]. Acc, Chem. Res. 2008, 41(6): 699-707.

[25] Huang J, Kaner R B. A general chemical route to polyaniline nanofibers[J]. J. Am. Chem. Soc, 2004, 126(3): 851-855.

[26] Huang J, Kaner R B. Nanofiber formation in the chemical polymerization of aniline: a mechanistic study[J]. Angew. Chem. Int. Ed., 2004, 116(43): 5817-5821.

[27] Ni W B, Wang D C, Huang Z J, et al. Fabrication of nanocomposite electrode

with MnO$_2$ nanoparticles distributed in polyaniline for electrochemical capacitors [J]. Mater. Chem. Phys., 2010, 124(2-3): 1151-1154.

[28] Wu Q, Xu Y, Yao Z, et al. Supercapacitors based on flexible graphene/polyaniline nanofiber composite films[J]. ACS Nano, 2010, 4(4): 1963-1970.

[29] (a) Wang Z L, Guo R, Li G R, et al. Polyaniline nanotube arrays as high-performance flexible electrodes for electrochemical energy storage devices[J]. J. Mater. Chem., 2012, 22(6): 2401-2404; (b) Boyd L W. Thin film growth by pulsed laser deposition[J]. Ceramics International, 1996, 22: 429-434.

[30] Yang G W, Xu C L, Li H L. Electrodeposited nickel hydroxide on nickel foam with ultrahigh capacitance[J]. Chem. Commun., 2008(48): 6537-6539.

[31] Zhao D, Zhou W, Li H. Effects of deposition potential and anneal temperature on the hexagonal nanoporous nickel hydroxide films[J]. Chem. Mater., 2007, 19(16): 3882-3891.

[32] Simon P, Gogotsi Y, Dunn B. Where do batteries end and supercapacitors begin? [J]. Science, 2014, 343: 1210-1211.

[33] Conway B E. Electrochemical supercapacitors[M]. New York: Kluwer-Plenum Pub. Co., 1999.

[34] Okubo M, Hosono E, Kim J, et al. Nanosize effect on high-Rate Li-Ion intercalation in LiCoO$_2$ electrode[J]. J. Am. Chem. Soc., 2007, 129(23): 7444-7452.

[35] Sathiya M, Prakash A S, Ramesha K, et al. V$_2$O$_5$-anchored carbon nanotubes for enhanced electrochemical energy storage[J]. J. Am. Chem. Soc., 2011, 133(40): 16291-16299.

[36] Augustyn V, Come J, Lowe M A, et al. High-rate electrochemical energy storage through Li$^+$ intercalation pseudocapacitance[J]. Nat. Mater., 2013, 12(6): 518-522.

[37] Brezesinski K, Wang J, Haetge J, et al. Pseudocapacitive contributions to charge storage in highly ordered mesoporous group V transition metal oxides with

iso-oriented layered nanocrystalline domains[J]. J. Am. Chem. Soc., 2010, 132(20): 6982-6990.

[38] Xie Y L, Li Z X, Xu Z G, et al. Preparation of coaxial $TiO_2$/ZnO nanotube arrays for high-efficiency photo-energy conversion applications[J]. Electrochem. Commun. 2011,13(8): 788-791.

[39] Kim T Y, Lee H W, Stoller M, et al. High-Performance supercapacitors based on poly(ionic liquid)-modified graphene electrodes[J]. ACS Nano, 2011, 5(1): 436-442.

[40] Wang H W, Hu Z A, Chang Y Q, et al. Facile solvothermal synthesis of a graphene nanosheet-bismuth oxide composite and its electrochemical characteristics[J]. Electrochim. Acta, 2010, 55(28): 8974-8980.

[41] Yoon S, Kang E, Kim J K, et al. Development of high-performance supercapacitor electrodes using novel ordered mesoporous tungsten oxide materials with high electrical conductivity[J]. Chem. Commun., 2011, 47(3): 1021-1023.

[42] Sun X, Wang G, Hwang J Y, et al. Porous nickel oxide nano-sheets for high performance pseudocapacitance materials[J]. J. Mater. Chem., 2011, 21(41): 16581-16588.

[43] Yuan C, Zhang X, Su L, et al. Facile synthesis and self-assembly of hierarchical porous NiO nano/micro spherical superstructures for high performance supercapacitors[J]. J. Mater. Chem., 2009, 19(32): 5772-5777.

[44] Lang J W, Kong L B, Wu W J, et al. Facile approach to prepare loose-packed NiO nano-flakes materials for supercapacitors[J]. Chem. Commun., 2008, (35): 4213-4215.

[45] Cao X, Shi Y, Shi W, et al. Preparation of novel 3D graphene networks for supercapacitor applications[J]. Small, 2011, 7(22): 3163-3168.

[46] Qiu Y J, Yu J, Zhou X S, et al. Synthesis of porous NiO and ZnO submicro-and nanofibers from electrospun polymer fiber templates[J]. Nanoscale Res. Lett.,

2009, 4: 173 - 177.

[47] Jiang J, Liu J, Zhou W, et al. CNT/Ni hybrid nanostructured arrays: synthesis and application as high-performance electrode materials for pseudocapacitors[J]. Energy Environ. Sci., 2011, 4(12): 5000 - 5007.

[48] Wu M S, Wang M J. Nickel oxide film with open macropores fabricated by surfactant-assisted anodic deposition for high capacitance supercapacitors[J]. Chem. Commun., 2010, 46(37): 6968 - 6970.

[49] Chidembo A, Aboutalebi S H, Konstantinov K, et al. Globular reduced graphene oxide-metal oxide structures for energy storage applications[J]. Energy Environ. Sci., 2012, 5(1): 5236 - 5240.

[50] Morant C, Soriano L, Trigo J F, et al. Atomic force microscope study of the early stages of NiO deposition on graphite and mica[J]. Thin Solid Films, 1998, 317(1 - 2): 59 - 63.

[51] Fujii E, Tomozawa A, Torii H, et al. Preferred orientations of NiO films prepared by plasma-enhanced metalorganic chemical vapor deposition[J]. Jpn. J. Appl. Phys., 1996, 35: L328 - L330.

[52] Lowndes D H, Geohegan D B, Puretzky A A, et al. Synthesis of novel thin-film materials by pulsed laser deposition[J]. Science, 1996, 273(5277): 898 - 903.

[53] Bae J, Song M K, Park Y J, et al. Fiber supercapacitors made of nanowire-fiber hybrid structures for wearable/flexible energy storage[J]. Angew. Chem. Int. Ed., 2011, 50(7): 1683 - 1687.

[54] Ji X, Lee K T, Nazar L F. A highly ordered nanostructured carbon-sulphur cathode for lithium-sulphur batteries[J]. Nat. Mater., 2009, 8: 500 - 506.

[55] Kim J Y, Lee K, Coates N E, et al. Efficient tandem polymer solar cells fabricated by all-solution processing[J]. Science, 2007, 317(5835): 222 - 225.

[56] Ormerod R M. Solid oxide fuel cells[J]. Chem. Soc. Rev., 2003, 32(1): 17 - 28.

[57] Su J, Guo L, Bao N, et al. Nanostructured $WO_3$/$BiVO_4$ heterojunction films for

efficient photoelectrochemical water splitting[J]. Nano Lett., 2011, 11(5): 1928-1933.

[58] Li R, Ren X, Zhang F, et al. Synthesis of $Fe_3O_4$@$SnO_2$ core-shell nanorod film and its application as a thin-film supercapacitor electrode[J]. Chem. Commun., 2012, 48(41): 5010-5012.

[59] Xu C, Zhao Y, Yang G, et al. Mesoporous nanowire array architecture of manganese dioxide for electrochemical capacitor applications[J]. Chem. Commun., 2009: 7575-7577.

[60] Zhou J, Liu J, Wang X D, et al. Vertically aligned $Zn_2SiO_4$ nanotube/ZnO nanowire heterojunction arrays[J]. Small, 2007, 3(4): 622-626.

[61] Xu C W, Wang H, Shen P K, et al. Highly ordered Pd nanowire arrays as effective electrocatalysts for ethanol oxidation in direct alcohol fuel cells[J]. Adv. Mater., 2007, 19(23): 4256-4259.

[62] Wu Z S, Wang D W, Ren W, et al. Anchoring hydrous $RuO_2$ on graphene sheets for high-performance electrochemical capacitors[J]. Adv. Funct. Mater., 2010, 20(20): 3595-3602.

[63] Wang H, Zhang L, Tan X, et al. Supercapacitive properties of hydrothermally synthesized $Co_3O_4$ nanostructures[J]. J. Phys. Chem. C, 2011, 115(35): 17599-17605.

[64] Lu X, Zheng D, Zhai T, et al. Facile synthesis of large-area manganese oxide nanorod arrays as a high-performance electrochemical supercapacitor[J]. Energy Environ. Sci., 2011, 4(8): 2915-2921.

[65] Yang D. Pulsed laser deposition of manganese oxide thin films for supercapacitor applications[J]. J. Power Sources, 2011, 196(20): 8843-8849.

[66] Julien C, Massot M, Baddour-Hadjean R, et al. Raman spectra of birnessite manganese dioxides[J]. Solid State Ionics, 2003, 159(3-4): 345-356.

[67] Zhang J, Jiang J, Zhao X S. Synthesis and capacitive properties of manganese oxide nanosheets dispersed on functionalized graphene sheets[J]. J. Phys.

Chem. C, 2011, 115(14): 6448 – 6454.

[68] Yuan L, Lu X H, Xiao X, et al. Flexible solid-state supercapacitors based on carbon nanoparticles/$MnO_2$ nanorods hybrid structure[J]. ACS Nano, 2012, 6(1): 656 – 661.

[69] Bao L, Zang J, Li X. Flexible $Zn_2SnO_4$/$MnO_2$ core/shell nanocable-carbon microfiber hybrid composites for high-performance supercapacitor electrodes[J]. Nano Lett., 2011, 11(3): 1215 – 1220.

[70] Lu Q, Lattanzi M W, Chen Y, et al. Supercapacitor electrodes with high-energy and power densities prepared from monolithic NiO/Ni nanocomposites[J]. Angew. Chem., 2011, 123(30): 6979 – 6982.

[71] Kung C W, Chen H W, Lin C Y, et al. Synthesis of $Co_3O_4$ nanosheets via electrodeposition followed by ozone treatment and their application to high-performance supercapacitors[J]. J. Power Sources, 2012, 214: 91 – 99.

[72] Simon P, Gogotsi Y. Materials for electrochemical capacitors[J]. Nat. Mater., 2008, 7: 845 – 854.

[73] Yu G, Hu L, Liu N, et al. Enhancing the supercapacitor performance of graphene/$MnO_2$ nanostructured electrodes by conductive wrapping[J]. Nano Lett., 2011, 11(10): 4438 – 4442.

[74] Wang G, Shen X, Horvat J, et al. Hydrothermal synthesis and optical, magnetic, and supercapacitance properties of nanoporous cobalt oxide nanorods[J]. J. Phys. Chem. C, 2009, 113(11): 4357 – 4361.

[75] Yan J, Sun W, We T, et al. Fabrication and electrochemical performances of hierarchical porous $Ni(OH)_2$ nanoflakes anchored on graphene sheets[J]. J. Mater. Chem., 2012, 22(23): 11494 – 11502.

[76] Chang S, Chen K, Hua Q. Evidence for the growth mechanisms of silver nanocubes and nanowires[J]. J. Phys. Chem. C, 2011, 115(16): 7979 – 7986.

[77] El-Kady M F, Strong V, Dubin S, et al. Laser scribing of high-performance and flexible graphene-based electrochemical capacitors [ J ]. Science, 2012,

335(6074): 1326-1330.

[78] Wang H, Wang Y, Wang X. Pulsed laser deposition of large-area manganese oxide nanosheet arrays for high-rate supercapacitors[J]. New J. Chem., 2013, 37(4) 869-872.

[79] Ionica-Bousquet C M, Casteel W J, Pearlstein R M, et al. Polyfluorinated boron cluster-[$B_{12}F_{11}H$]$^{2-}$-based electrolytes for supercapacitors: overcharge protection[J]. Electrochem. Commun., 2010, 12(5): 636-639.

[80] Liu H J, Cui W J, Jin L H, et al. Preparation of three-dimensional ordered mesoporous carbon sphere arrays by a two-step templating route and their application for supercapacitors[J]. J. Mater. Chem., 2009, 19(22): 3661-3667.

[81] Frackowiak E. Carbon materials for supercapacitor application[J]. Phys. Chem. Chem. Phys., 2007, 9(15): 1774-1785.

[82] Tang Z, Tang C, Gong H. A high energy density asymmetric supercapacitor from nano-architectured $Ni(OH)_2$/carbon nanotube electrodes[J]. Adv. Funct. Mater., 2012, 22(6): 1272-1278.

[83] Lu X, Yu M, Wang G, et al. H-$TiO_2$@$MnO_2$//H-$TiO_2$@C core-shell nanowires for high performance and flexible asymmetric supercapacitors[J]. Adv. Mater., 2013, 25(2): 267-272.

[84] Hu C C, Chen J C, Chang K H. Cathodic deposition of $Ni(OH)_2$ and $Co(OH)_2$ for asymmetric supercapacitors: Importance of the electrochemical reversibility of redox couples[J]. J. Power Sources, 2013, 221, 128-133.

[85] Lu X, Yu M, Zhai T, et al. High energy density asymmetric quasi-solid-state supercapacitor based on porous vanadium nitride nanowire anode[J]. Nano Lett., 2013, 13(6): 2628-2633.

[86] Yan J, Fan Z, Sun W, et al. Advanced asymmetric supercapacitors based on $Ni(OH)_2$/graphene and porous graphene electrodes with high energy density[J]. Adv. Funct. Mater., 2012, 22(12): 2632-2641.

[87] Nam K W, Yoon W S, Kim K B. X-ray absorption spectroscopy studies of nickel oxide thin film electrodes for supercapacitors[J]. Electrochim. Acta, 2002, 47(19): 3201-3209.

[88] Luan F, Wang G, Ling Y, et al. High energy density asymmetric supercapacitors with a nickel oxide nanoflake cathode and a 3D reduced graphene oxide anode[J]. Nanoscale, 2013, 5(17): 7984-7990.

[89] Zhao D D, Xu M W, Zhou W J, et al. Preparation of ordered mesoporous nickel oxide film electrodes via lyotropic liquid crystal templated electrodeposition route [J]. Electrochim. Acta 2008, 53(6): 2699-2705.

[90] Wang D W, Li F, Cheng H M. Hierarchical porous nickel oxide and carbon as electrode materials for asymmetric supercapacitor[J]. J. Power Sources, 2008, 185(2): 1563-1568.

[91] Jiang J, Li Y, Liu J, et al. Recent advances in metal oxide-based electrode architecture design for electrochemical energy storage[J]. Adv. Mater., 2012, 24(38): 5166-5180.

[92] Kim J Y, Lee S H, Yan Y, et al. Controlled synthesis of aligned Ni-NiO core-shell nanowire arrays on glass substrates as a new supercapacitor electrode[J]. RSC Adv., 2012, 2(22): 8281-8285.

[93] Lv W, Sun F, Tang D M, et al. A sandwich structure of graphene and nickel oxide with excellent supercapacitive performance[J]. J. Mater. Chem., 2011, 21(25): 9014-9019.

[94] Yang Y Y, Hu Z A, Zhang Z Y, et al. Reduced graphene oxide-nickel oxide composites with high electrochemical capacitive performance[J]. Mater. Chem. Phys., 2012, 133(1): 363-368.

[95] Wu M S, Lin Y P, Lin C H, et al. Formation of nano-scaled crevices and spacers in NiO-attached graphene oxide nanosheets for supercapacitors[J]. J. Mater. Chem., 2012, 22(6): 2442-2448.

[96] Zhao B, Song J, Liu P, et al. Monolayer graphene/NiO nanosheets with two-

dimension structure for supercapacitors[J]. J. Mater. Chem., 2011, 21(46): 18792-18798.

[97] Chen Z, Ren W, Gao L, et al. Cheng, Three-dimensional flexible and conductive interconnected graphene networks grown by chemical vapour deposition[J]. Nat. Mater., 2011, 10(6): 424-428.

[98] Dong X C, Xu H, Wang X W, et al. 3D graphene-cobalt oxide electrode for high-performance supercapacitor and enzymeless glucose detection[J]. ACS Nano, 2012, 6(4): 3206-3213.

[99] Ji H, Zhang L, Pettes M T, et al. Ultrathin graphite foam: a three-dimensional conductive network for battery electrodes[J]. Nano Lett., 2012, 12(5): 2446-2451.

[100] Li H B, Yu M H, Wang F X, et al. Amorphous nickel hydroxide nanospheres with ultrahigh capacitance and energy density as electrochemical pseudocapacitor materials[J]. Nat. Commun., 2013, 4, 1894-1900.

[101] Qie L, Chen W M, Wang Z H, et al. Nitrogen-doped porous carbon nanofiber webs as anodes for lithium ion batteries with a superhigh capacity and rate capability[J]. Adv. Mater., 2012, 24(15) 2047-2450.

[102] Qie L, Chen W, Xu H, et al. Synthesis of functionalized 3D hierarchical porous carbon for high-performance supercapacitors[J]. Energy Environ. Sci., 2013, 6(8): 2497-2504.

[103] Dai T, Lu Y. Water-soluble methyl orange fibrils as versatile templates for the fabrication of conducting polymer microtubules [J]. Macromol. Rapid Commun., 2007, 28(5): 629-633.

[104] Yoon H, Chang M, Jang J. Sensing behaviors of polypyrrole nanotubes prepared in reverse microemulsions: effects of transducer size and transduction mechanism[J]. J. Phys. Chem. B, 2006, 110(29): 14074-14077.

[105] Wang Y, Wang H, Wang X. The cobalt oxide/hydroxide nanowall array film prepared by pulsed laser deposition for supercapacitors with superb-rate

capability[J]. Electrochim. Acta, 2013, 92: 298 – 303.

[106] Meyer J C, Geim A K, Katsnelson M I, et al. The structure of suspended graphene sheets[J]. Nature, 2007, 446: 60 – 63.

[107] He Y, Chen W, Li X, Zhang Z, et al. Freestanding three-dimensional graphene/$MnO_2$ composite networks as ultralight and flexible supercapacitor electrodes[J]. ACS Nano, 2013, 7(1): 174 – 182.

[108] Wang W, Guo S, Penchev M, et al. Three dimensional few layer graphene and carbon nanotube foam architectures for high fidelity supercapacitors[J]. Nano Energy, 2013, 2(2): 294 – 303.

[109] Wang H, Yi H, Chen X, et al. Facile synthesis of a nano-structured nickel oxide electrode with outstanding pseudocapacitive properties[J]. Electrochim. Acta, 2013, 105: 353 – 361.

[110] Zhou W, Cao X, Zeng Z, et al. One-step synthesis of $Ni_3S_2$ nanorod@ $Ni(OH)_2$ nanosheet core-shell nanostructures on a three-dimensional graphene network for high-performance supercapacitors[J]. Energy Environ. Sci., 2013, 6(7): 2216 – 2221.

[111] Zhang X, Shi W, Zhu J, et al. Synthesis of porous NiO nanocrystals with controllable surface area and their application as supercapacitor electrodes[J]. Nano Res. 2010, 3(9): 643 – 652.

[112] Wang B, Chen J S, Wang Z, et al. Green synthesis of NiO nanobelts with exceptional pseudo-capacitive properties [J]. Adv. Energy Mater., 2012, 2(10): 1188 – 1192.

[113] Zhang L L, Zhao X, Stoller M D, et al. Highly conductive and porous activated reduced graphene oxide films for high-power supercapacitors[J]. Nano Lett., 2012, 12(4): 1806 – 1812.

[114] Chen L F, Zhang X D, Liang H W, et al. Synthesis of nitrogen-doped porous carbon nanofibers as an efficient electrode material for supercapacitors[J]. ACS Nano, 2012, 6(8): 7092 – 7102.

[115] Wu Z S, Ren W, Wang D W, et al. High-energy $MnO_2$ nanowire/graphene and graphene asymmetric electrochemical capacitors[J]. ACS Nano, 2010, 4(10): 5835–5842.

[116] Zhuiykov S, Kats E. Atomically thin two-dimensional materials for functional electrodes of electrochemical devices[J]. Ionics, 2013, 19: 825–865.

[117] Wang X, Liu W S, Lu X H, et al. Dodecyl sulfate-induced fast faradic process in nickel cobalt oxide-reduced graphite oxide composite material and its application for asymmetric supercapacitor device[J]. J. Mater. Chem., 2012, 22(43): 23114–23119.

[118] Tang C H, Tang Z, Gong H. Hierarchically porous Ni–Co oxide for high reversibility asymmetric full-cell supercapacitors[J]. J. Electrochem. Soc., 2012, 159(5): A651–A656.

[119] Wang H, Liang Y, Mirfakhrai T, et al. Advanced asymmetrical supercapacitors based on graphene hybrid materials[J]. Nano Res., 2011, 4(8): 729–736.

[120] Zhou C, Zhang Y, Li Y, et al. Construction of high-capacitance 3D CoO@polypyrrole nanowire array electrode for aqueous asymmetric supercapacitor[J]. Nano Lett., 2013, 13(5): 2078–2085.

[121] Wang H, Holt C M B, Li Z, et al. Graphene-nickel cobaltite nanocomposite asymmetrical supercapacitor with commercial level mass loading[J]. Nano Res., 2012, 5(9): 605–617.

[122] Fan Z, Yan J, Wei T, et al. Asymmetric supercapacitors based on graphene/$MnO_2$ and activated carbon nanofiber electrodes with high power and energy density[J]. Adv. Funct. Mater., 2011, 21(12): 2366–2375.

[123] Zhang J, Jiang J, Li H, et al. A high-performance asymmetric supercapacitor fabricated with graphene-based electrodes[J]. Energy Environ. Sci., 2011, 4(10): 4009–4015.

[124] Croce F, Appetecchi G B, Persi L, et al. Nanocomposite polymer electrolytes

for lithium batteries[J]. Nature, 1998, 394: 456 – 458.

[125] Nelson R F. Power requirements for batteries in hybrid electric vehicles[J]. J. Power Sources, 2000, 91(1): 2 – 26.

[126] Zhou G, Wang D W, Li F, et al. Graphene-wrapped $Fe_3O_4$ anode material with improved reversible capacity and cyclic stability for lithium ion batteries[J]. Chem. Mater., 2010, 22(18): 5306 – 5313.

[127] Yuan G, Jiang Z, Aramat A, et al. Electrochemical behavior of activated-carbon capacitor material loaded with nickel oxide[J]. Carbon, 2005, 43(14): 2913 – 2917.

[128] Sharma Y, Sharma N, Subba Rao G V, et al. Nanophase $ZnCo_2O_4$ as a high performance anode material for Li-ion batteries[J]. Adv. Funct. Mater. 2007, 17(15): 2855 – 2861.

[129] Wang Q, Liu B, Wang X, et al. Morphology evolution of urchin-like $NiCo_2O_4$ nanostructures and their applications as psuedocapacitors and photoelectrochemical cells[J]. J. Mater. Chem., 2012, 22: 21647 – 21653.

[130] Zhu J, Gao Q. Mesoporous $MCo_2O_4$ (M =Cu, Mn and Ni) spinels: structural replication, characterization and catalytic application in CO oxidation [J]. Microporous Mesoporous Mater., 2009, 124(1 – 3): 144 – 152.

[131] Liang Y, Wang H, Zhou J, et al. Oxygen reduction electrocatalyst based on strongly coupled cobalt oxide nanocrystals and carbon nanotubes[J]. J. Am. Chem. Soc., 2012, 134(38): 15849 – 15857.

[132] Wang H, Yang Y, Liang Y, et al. Rechargeable Li – $O_2$ batteries with a covalently coupled $MnCo_2O_4$-graphene hybrid as an oxygen cathode catalyst[J]. Energy Environ. Sci., 2012, 5(7): 7931 – 7935.

[133] Wang H, Gao Q, Jiang L. Facile approach to prepare nickel cobaltite nanowire materials for supercapacitors[J]. Small, 2011, 7(17): 2454 – 2459.

[134] Wang X, Han X, Lim M, et al. Nickel cobalt oxide-single wall carbon nanotube composite material for superior cycling stability and high-performance

supercapacitor application [J]. J. Phys. Chem. C, 2012, 116 (23): 12448-12454.

[135] Ding R, Qi L, Wang H. A facile and cost-effective synthesis of mesoporous NiCo$_2$O$_4$ nanoparticles and their capacitive behavior in electrochemical capacitors [J]. J Solid State Electrochem. 2012, 16, 3621-3633.

[136] Chang S K, Lee K T, Zainal Z, et al. Structural and electrochemical properties of manganese substituted nickel cobaltite for supercapacitor application [J]. Electrochim. Acta, 2012, 67: 67-72.

[137] Wang H W, Hu Z A, Chang Y Q, et al. Design and synthesis of NiCo$_2$O$_4$-reduced graphene oxide composites for high performance supercapacitors [J]. J. Mater. Chem., 2011, 21(28): 10504-10511.

[138] Yuan C, Li J, Hou L, et al. Facile template-free synthesis of ultralayered mesoporous nickel cobaltite nanowires towards high-performance electrochemical capacitors [J]. J. Mater. Chem., 2012, 22 (31): 16084-16090.

[139] Jiang H, Ma J, Li C. Hierarchical porous NiCo$_2$O$_4$ nanowires for high-rate supercapacitors[J]. Chem. Commun., 2012, 48(37): 4465-4467.

[140] Lu Q, Chen Y, Li W, et al. Ordered mesoporous nickel cobaltite spinel with ultra-high supercapacitance[J]. J. Mater. Chem. A, 2013, 1(6): 2331-2336.

[141] Wu T, Li J, Hou L, et al. Uniform urchin-like nickel cobaltite microspherical superstructures constructed by one-dimension nanowires and their application for electrochemical capacitors[J]. Electrochim. Acta, 2012, 81, 172-178.

[142] Xiao J, Yang S. Sequential crystallization of sea urchin-like bimetallic (Ni, Co) carbonate hydroxide and its morphology conserved conversion to porous NiCo$_2$O$_4$ spinel for pseudocapacitors[J]. RSC Adv. 2011, 1(4): 588-595.

[143] Chien H C, Cheng W Y, Wang Y H, et al. Ultrahigh specific capacitances for supercapacitors achieved by nickel cobaltite/carbon aerogel composites [J]. Adv. Funct. Mater., 2012, 22(23): 5038-5043.

[144] Zhang G Q, Wu H B, Hoster H E, et al. Single-crystalline $NiCo_2O_4$ nanoneedle arrays grown on conductive substrates as binder-free electrodes for high-performance supercapacitors[J]. Energy Environ. Sci., 2012, 5(11): 9453-9456.

[145] Wang Q, Wang X, Liu B, et al. $NiCo_2O_4$ nanowire arrays supported on Ni foam for high-performance flexible all-solid-state supercapacitors[J]. J. Mater. Chem. A, 2013, 1(7): 2468-2473.

[146] Yuan C, Li J, Hou L, et al. Ultrathin mesoporous $NiCo_2O_4$ nanosheets supported on Ni foam as advanced electrodes for supercapacitors[J]. Adv. Funct. Mater. 2012, 22(21): 4592-4597.

[147] Zhang G, Lou X W. General solution growth of mesoporous $NiCo_2O_4$ nanosheets on various conductive substrates as high-performance electrodes for supercapacitors[J]. Adv. Mater. 2013, 25(7): 976-979.

[148] Xing W, Qiao S Z, Wu X Z, et al. Exaggerated capacitance using electrochemically active nickel foam as current collector in electrochemical measurement[J]. J. Power Sources, 2011, 196(8): 4123-4127.

[149] Grdeń M, Alsabet M, Jerkiewicz G. Surface science and electrochemical analysis of nickel foams[J]. ACS Appl. Mater. Interfaces, 2012, 4(6): 3012-3021.

[150] Rakhi R B, Chen W, Cha D, et al. Substrate dependent self-organization of mesoporous cobalt oxide nanowires with remarkable pseudocapacitance[J]. Nano Lett., 2012, 12(5): 2559-2567.

[151] Gupta V, Gupta S, Miura N. Electrochemically synthesized nanocrystalline spinel thin film for high performance supercapacitor[J]. J. Power Sources, 2010, 195(11): 3757-3760.

[152] Kim J H, Lee K H, Overzet L J, et al. Synthesis and electrochemical properties of spin-capable carbon nanotube sheet/$MnO_x$ composites for high-performance energy storage devices [J]. Nano Lett., 2011, 11(7):

2611-2617.

[153] Zhang X, Gong L, Liu K, et al. Tungsten oxide nanowires grown on carbon cloth as a flexible cold cathode[J]. Adv. Mater., 2010, 22(46): 5292-5296.

[154] Liu B, Zhang J, Wang X, et al. Hierarchical three-dimensional $ZnCo_2O_4$ nanowire arrays/carbon cloth anodes for a novel class of high-performance flexible lithium-ion batteries[J]. Nano Lett., 2012, 12(6): 3005-3011.

[155] Wang Z, Wang H, Liu B, et al. Transferable and flexible nanorod-assembled $TiO_2$ cloths for dye-sensitized solar cells, photodetectors, and photocatalysts [J]. ACS Nano, 2011, 5(10): 8412-8419.

[156] Liu B, Wang Z, Dong Y, et al. ZnO-nanoparticle-assembled cloth for flexible photodetectors and recyclable photocatalysts[J]. J. Mater. Chem., 2012, 22(18): 9379-9384.

[157] Wang H, Wang X. Growing nickel cobaltite nanowires and nanosheets on carbon cloth with different pseudocapacitive performance[J]. ACS Appl. Mater. Interfaces, 2013, 5(13): 6255-6260.

[158] Jiang J, Liu J P, Huang X T, et al. General synthesis of large-scale arrays of one-dimensional nanostructured $Co_3O_4$ directly on heterogeneous substrates[J]. Cryst. Growth Des., 2010, 10(1): 70-75.

[159] Liu J, Jiang J, Cheng C, et al. $Co_3O_4$ nanowire@$MnO_2$ ultrathin nanosheet core/shell arrays: a new class of high-performance pseudocapacitive materials [J]. Adv. Mater. 2011, 23(18): 2076-2081.

[160] Xia X, Tu J, Zhang Y, et al. High-quality metal oxide core/shell nanowire arrays on conductive substrates for electrochemical energy storage[J]. ACS Nano, 2012, 6(6): 5531-5538.

[161] Tao Y S, Kanoh H, Abrams L, et al. Mesopore-modified zeolites: preparation, characterization, and applications [J]. Chem. Rev., 2006, 106(3): 896-910.

[162] Yoon S M, Choi W M, Baik H, et al. Synthesis of multilayer graphene balls by

carbon segregation from nickel nanoparticles[J]. ACS Nano, 2012, 6(8): 6803-6811.

[163] Yu J G, Wang G H, Cheng B, et al. Effects of hydrothermal temperature and time on the photocatalytic activity and microstructures of bimodal mesoporous $TiO_2$ powders[J]. Appl. Catal. B, 2007, 69(3-4): 171-180.

[164] Xia X, Tu J, Mai Y, et al. Graphene sheet/porous NiO hybrid film for supercapacitor applications[J]. Chem. Eur. J., 2011, 17(39): 10898-10905.

[165] Wang H W, Hu Z A, Chang Y Q, et al. Preparation of reduced graphene oxide/cobalt oxide composites and their enhanced capacitive behaviors by homogeneous incorporation of reduced graphene oxide sheets in cobalt oxide matrix[J]. Mater. Chem. Phys. 2011, 130(1-2): 672-679.

[166] Li J, Xiong S, Liu Y, et al. High electrochemical performance of monodisperse $NiCo_2O_4$ mesoporous microspheres as an anode material for Li-ion batteries[J]. ACS Appl. Mater. Interfaces, 2013, 5(3): 981-988.

[167] Wu Y P, Wang F, Xiao S, et al. Electrode materials for aqueous asymmetric supercapacitors[J]. RSC Adv., 2013, 3(32): 13059-13084.

[168] Liang Y, Wang H, Sanchez H, Casalongue, et al. $TiO_2$ nanocrystals grown on graphene as advanced photocatalytic hybrid materials[J]. Nano Res., 2010, 3(10): 701-705.

[169] Wang H, Wang Y, Hu Z, Wang X. Cutting and unzipping multiwalled carbon nanotubes into curved graphene nanosheets and their enhanced supercapacitor performance[J]. ACS Appl. Mater. Interfaces, 2012, 4(12): 6827-6834.

[170] Xie K, Qin X, Wang X, et al. Carbon nanocages as supercapacitor electrode materials[J]. Adv. Mater., 2012, 24(3): 347-352.

[171] Guan C, Li X, Wang Z, et al. Nanoporous walls on macroporous foam: rational design of electrodes to push areal pseudocapacitance[J]. Adv. Mater., 2012, 24(30): 4186-4190.

[172] Chen Y L, Hu Z A, Chang Y Q, et al. Zinc oxide/reduced graphene oxide

composites and electrochemical capacitance enhanced by homogeneous incorporation of reduced graphene oxide sheets in zinc oxide matrix[J]. J. Phys. Chem. C, 2011, 115(5): 2563-2571.

[173] Izadi-Najafabadi A, Yamada T, Futaba D N, et al. High-power supercapacitor Electrodes from Single-Walled Carbon Nanohorn/Nanotube Composite[J]. ACS Nano, 2011, 5(2): 811-819.

[174] Kay H An, Won S K, Young S P, et al. Supercapacitors using single-walled carbon nanotube electrodes[J]. Adv. Mater. 2001, 13(7): 497-500.

[175] Yang C M, Kim Y J, Endo M, et al. Nanowindow-regulated specific capacitance of supercapacitor electrodes of single-wall carbon nanohorns[J]. J. Am. Chem. Soc., 2007, 129(1): 20-21.

[176] Frackowiak E, Béguin F. Carbon materials for the electrochemical storage of energy in capacitors[J]. Carbon, 2001, 39(6): 937-950.

[177] Rubio N, Fabbro C, Herrero M A, et al. Ball-milling modification of single-walled carbon nanotubes: purification, cutting, and functionalization [J]. Small, 2011, 7(5): 665-674.

[178] Lee J, Jeong T, Heo J, et al. Short carbon nanotubes produced by cryogenic crushing[J]. Carbon, 2006, 44(14): 2984-2989.

[179] Wang X X, Wang J N. Preparation of short and water-dispersible carbon nanotubes by solid-state cutting[J]. Carbon, 2008, 46(1): 117-125.

[180] Wang X X, Wang J N, Su L F, et al. Cutting of multi-walled carbon nanotubes by solid-state reaction[J]. J. Mater. Chem., 2006, 16(43): 4231-4234.

[181] Wang Y, Zhang J, Zang J, et al. Etching and cutting of multi-walled carbon nanotubes in molten nitrate[J]. Corrosion Science, 2011, 53(11): 3764-3770.

[182] Gu Z, Peng H, Hauge R H, et al. Cutting single-wall carbon nanotubes through fluorination[J]. Nano Lett., 2002, 2(9): 1009-1013.

[183] Ziegler K J, Gu Z, Peng H, et al. Controlled oxidative cutting of single-walled carbon nanotubes[J]. J. Am. Chem. Soc., 2005, 127(5): 1541-1547.

[184] Shinde D B, Pillai V K. Electrochemical preparation of luminescent graphene quantum dots from multiwalled carbon nanotubes[J]. Chem. Eur. J., 2012, 18(39): 12522-12528.

[185] Wang X X, Wang J N, Chang H, et al. Preparation of short carbon nanotubes and application as an electrode material in Li-ion batteries[J]. Adv. Funct. Mater., 2007, 17(17): 3613-3618.

[186] Kalita G, Adhikari S, Aryal H R, et al. Fullerene ($C_{60}$) decoration in oxygen plasma treated multiwalled carbon nanotubes for photovoltaic application[J]. Appl. Phys. Lett., 2008, 92(6): 123508-123511.

[187] Jiao L, Zhang L, Wang X, et al. Narrow graphene nanoribbons from carbon nanotubes[J]. Nature, 2009, 458: 877-880.

[188] Kosynkin D V, Higginbotham A L, Sinitskii A, et al. Longitudinal unzipping of carbon nanotubes to form graphene nanoribbons[J]. Nature, 2009, 458: 872-876.

[189] Wang Y, Shi Z X, Yin J. Unzipped multiwalled carbon nanotubes for mechanical reinforcement of polymer composites[J]. J. Phys. Chem. C, 2010, 114(46): 19621-19628.

[190] Shinde D B, Debgupta J, Kushwaha A, et al. Electrochemical unzipping of Multi-walled carbon nanotubes for facile synthesis of High-quality graphene nanoribbons[J]. J. Am. Chem. Soc., 2011, 133(12): 4168-4171.

[191] Kosynkin D V, Lu W, Sinitskii A, et al. Highly conductive graphene nanoribbons by longitudinal splitting of carbon nanotubes using potassium vapor[J]. ACS Nano, 2011, 5(2): 968-974.

[192] Elías A L, Botello-Méndez A R, Meneses-Rodríguez D, et al. Longitudinal cutting of pure and doped carbon nanotubes to form graphitic nanoribbons using metal clusters as nanoscalpels[J]. Nano Lett., 2010, 10(2): 366-372.

[193] Morelos-Gómez A M, Vega-Díaz S, González V J, et al. Clean nanotube unzipping by abrupt thermal expansion of molecular nitrogen: graphene

nanoribbons with atomically smooth edges[J]. ACS Nano, 2012, 6(3): 2261-2272.

[194] Fang W C, Huang J H, Chen L C, et al. Carbon nanotubes grown directly on Ti electrodes and enhancement of their electrochemical properties by nitric acid treatment[J]. Electrochem. Solid-State Lett., 2006, 9(1): 5-8.

[195] Long D, Li W, Qiao W, et al. Partially unzipped carbon nanotubes as a superior catalyst support for PEM fuel cells[J]. Chem. Commun., 2011, 47(33): 9429-9431.

[196] Wang G, Ling Y, Qian F, et al. Enhanced capacitance in partially exfoliated multi-walled carbon nanotubes[J]. J. Power Sources, 2011, 196(11): 5209-5214.

[197] Hummers W S, Offeman R E. Preparation of graphitic oxide[J]. J. Am. Chem. Soc., 1958, 80: 1339.

[198] Chang J, Sun J, Xu C, et al. Template-free approach to synthesize hierarchical porous nickel cobalt oxides for supercapacitors[J]. Nanoscale, 2012, 4(21): 6786-6791.

[199] Stankovich S, Piner R D, Chen X, et al. Stable aqueous dispersions of graphitic nanoplatelets via the reduction of exfoliated graphite oxide in the presence of poly(sodium 4-styrenesulfonate)[J]. J. Mater. Chem., 2006, 16(2): 155-158.

[200] Stankovich S, Dikin D A, Piner R D, et al. Synthesis of graphene-based nanosheets via chemical reduction of exfoliated graphite oxide[J]. Carbon, 2007, 45(7): 1558-1565.

[201] Wang Y, Shi Z X, Yin J. Kevlar oligomer functionalized graphene for polymer composites[J]. Polymer, 2011, 52(16): 3661-3670.

[202] Wang Y, Shi Z X, Yin J. Facile synthesis of soluble graphene via a green reduction of graphene oxide in tea solution and its biocomposites[J]. ACS Appl. Mater. Interfaces, 2011, 3(4): 1127-1133.

[203] Dreyer D R, Park S, Bielawski C W, et al. The chemistry of graphene oxide [J]. Chem. Soc. Rev., 2010, 39(1): 228-240.

[204] Dreyer D R, Ruoff R S, Bielawski C W. From conception to realization: an historial account of graphene and some perspectives for its future[J]. Angew. Chem. Int. Ed., 2010, 49(49): 9336-9344.

[205] Hirsch A. Unzipping carbon nanotubes: a peeling method for the formation of graphene nanoribbons [J]. Angew. Chem. Int. Ed., 2009, 48 (36): 6594-6596.

[206] Olley R H, Bassett D C. An improved permanganic etchant for polyolefines[J]. Polymer, 1982, 23(12): 1707-1710.

[207] Zhao X, Chu B T T, Ballesteros B, et al. Spray deposition of steam treated and functionalized single-walled and multi-walled carbon nanotube films for supercapacitors[J]. Nanotechnology, 2009, 20(6): 065605-065613.

[208] Zhong X, H. Li Y L, Liu Y K, et al. Continuous multilayered carbon nanotube yarns[J]. Adv. Mater., 2010, 22(6): 692-696.

[209] Masarapu C, Zeng H F, Hung K H, et al. Effect of temperature on the Capacitance of Carbon Nanotube Supercapacitors[J]. ACS Nano, 2009, 3(8): 2199-2206.

[210] Li X, Rong J, Wei B. Electrochemical behavior of single-walled carbon nanotube supercapacitors under compressive stress[J]. ACS Nano, 2010, 4(10): 6039-6049.

[211] An K H, Kim W S, Park Y S, et al. Electrochemical properties of high-power supercapacitors using single-walled carbon nanotube electrodes [J]. Adv. Funct. Mater., 2001, 11: 387-392.

[212] Kim J H, Nam K W, Ma S B, et al. Fabrication and electrochemical properties of carbon nanotube film electrodes[J]. Carbon, 2006, 44(10): 1963-1968.

[213] Du C, Yeh J, Pan N. High power density supercapacitors using locally aligned carbon nanotube electrodes[J]. Nanotechnology, 2005, 16(4): 350-353.

[214] Jurewicz K, Babeł K, Pietrzak R, et al. Capacitance properties of multi-walled carbon nanotubes modified by activation and ammoxidation[J]. Carbon, 2006, 44(12): 2368-2375.

[215] Jang I Y, Muramatsu H, Park K C, et al. Capacitance response of double-walled carbon nanotubes depending on surface modification[J]. Electrochem. Commun., 2009, 11(4): 719-723.

[216] Niu Z, Zhou W, Chen J, et al. Compact-designed supercapacitors using free-standing single-walled carbon nanotube films[J]. Energy Environ. Sci., 2011, 4(4): 1440-1446.

[217] Lee S W, Kim B S, Chen S, et al. Layer-by-layer assembly of all carbon nanotube ultrathin films for electrochemical applications[J]. J. Am. Chem. Soc., 2009, 131(2): 671-679.

[218] Kaempgen M, Chan C K, Ma J, et al. Printable thin film supercapacitors using single-walled carbon nanotubes[J]. Nano Lett., 2009, 9(5): 1872-1876.

[219] Hiraoka T, Izadi-Najafabadi A, Yamada T, et al. Compact and light supercapacitor electrodes from a surface-only solid by opened carbon nanotubes with 2200 $m^2 \cdot g^{-1}$ surface area[J]. Adv. Funct. Mater., 2010, 20(3): 422-428.

[220] Hu L, Choi J W, Yang Y, et al. Highly conductive paper for energy-storage devices[J]. PNAS, 2009, 106(51): 21490-21494.

[221] Gowda S R, Reddy A L M, Zhan X, et al. Building energy storage device on a single nanowire[J]. Nano Lett., 2011, 11(8): 3329-3333.

[222] Xiang Q, Yu J, Jaroniec M. Preparation and enhanced visible-light photocatalytic $H_2$-production activity of graphene/$C_3N_4$ composites[J]. J. Phys. Chem. C, 2011, 115(15): 7355-7363.

[223] Gao B, Yuan C, Su L, et al. High dispersion and electrochemical capacitive performance of NiO on benzenesulfonic functionalized carbon nanotubes[J]. Electrochim. Acta, 2009, 54(13): 3561-3567.

[224] Xu C, Wang X, Zhu J. Graphene-metal particle nanocomposites[J]. J. Phys. Chem. C, 2008, 112(50): 19841-19845.

[225] Xu J, Wang Q, Wang X, et al. Flexible asymmetric supercapacitors based upon $Co_9S_8$ nanorod//$Co_3O_4$@$RuO_2$ nanosheet arrays on carbon cloth[J]. ACS Nano, 2013, 7(6): 5453-5462.

[226] Lin T, Huang F, Liang J, et al. A facile preparation route for boron-doped graphene, and its CdTe solar cell application[J]. Energy Environ. Sci., 2011, 4(3): 862-865.

[227] Xia X, Tu J, Zhang Y, et al. Porous hydroxide nanosheets on preformed nanowires by electrodeposition: branched nanoarrays for electrochemical energy storage[J]. Chem. Mater., 2012, 24(19): 3793-3799.

[228] Deng D, Pan X, Zhang H, et al. Freestanding graphene by thermal splitting of silicon carbide granules[J]. Adv. Mater., 2010, 22(19): 2168-2171.

[229] Yuan C, Yang L, Hou L, et al. Growth of ultrathin mesoporous $Co_3O_4$ nanosheet arrays on Ni foam for high-performance electrochemical capacitors [J]. Energy Environ. Sci., 2012, 5(7): 7883-7887.

[230] Huang L, Chen D, Ding Y, et al. Nickel-cobalt hydroxide nanosheets coated on $NiCo_2O_4$ nanowires grown on carbon fiber paper for high-performance pseudocapacitors[J]. Nano Lett., 2013, 13(7): 3135-3139.

[231] Shao M, Ning F, Zhao Y, et al. Core-shell layered double hydroxide microspheres with tunable interior architecture for supercapacitors[J]. Chem. Mater., 2012, 24(6): 1192-1197.

[232] Li Q, Wang Z L, Li G R, et al. Design and synthesis of $MnO_2$/Mn/$MnO_2$ sandwich-structured nanotube arrays with high supercapacitive performance for electrochemical energy Storage[J]. Nano Lett., 2012, 12(7): 3803-3807.

[233] Du C, Pan N. Supercapacitors using carbon nanotubes films by electrophoretic deposition[J]. J. Power Sources, 2006, 160(2): 1487-1494.

[234] Xiao W, Xia H, Fuh J Y H, et al. Growth of single-crystal α-$MnO_2$

nanotubes prepared by a hydrothermal route and their electrochemical properties [J]. J. Power Sources, 2009, 193(2): 935-938.

[235] Perera S D, Patel B, Nijem N, et al. Vanadium oxide nanowire-carbon nanotube binder-free flexible electrodes for supercapacitors[J]. Adv. Energy Mater., 2011, 1(5): 936-945.

[236] Biswas S, Drzal L T. Multilayered nanoarchitecture of graphene nanosheets and polypyrrole nanowires for high performance supercapacitor electrodes[J]. Chem. Mater., 2010, 22(20): 5667-5671.

[237] Feng J, Sun X, Wu C, et al. Metallic few-layered $VS_2$ ultrathin nanosheets: high two-dimensional conductivity for in-plane supercapacitors[J]. J. Am. Chem. Soc., 2011, 133(44): 17832-17838.

[238] Zhong J H, Wang A L, Li G R, et al. $Co_3O_4/Ni(OH)_2$ composite mesoporous nanosheet networks as a promising electrode for supercapacitor applications[J]. J. Mater. Chem., 2012, 22(12): 5656-5665.

[239] Lu Z, Chang Z, Zhu W, et al. Beta-phased $Ni(OH)_2$ nanowall film with reversible capacitance higher than theoretical Faradic capacitance[J]. Chem. Commun., 2011, 47(34): 9651-9653.

[240] Bi R R, Wu X L, Cao F F, et al. Highly dispersed $RuO_2$ nanoparticles on carbon nanotubes: facile synthesis and enhanced supercapacitance performance [J]. J. Phys. Chem. C, 2010, 114(6): 2448-2451.

[241] Chen L L, Wu X L, Guo Y G, et al. Synthesis of nanostructured fibers consisting of carbon coated $Mn_3O_4$ nanoparticles and their application in electrochemical capacitors [J]. J. Nanosci. Nanotechnol., 2010, 10(12): 8158-8163.

[242] Dikin D A, Stankovich S, Zimney E J, et al. Preparation and characterization of graphene oxide paper[J]. Nature, 2007, 448(7152): 457-460.

[243] Chen Z, Ren W, Gao L, et al. Three-dimensional flexible and conductive interconnected graphene networks grown by chemical vapour deposition[J].

Nat. Mater., 2011, 10(6): 424-428.

[244] Fan Z, Yan J, Zhi L, et al. A three-dimensional carbon nanotube/graphene sandwich and its application as electrode in supercapacitors[J]. Adv. Mater., 2010, 22(33): 3723-3728.

[245] Sinitskii A, Dimiev A, Kosynkin D V, et al. Graphene nanoribbon devices produced by oxidative unzipping of carbon nanotubes[J]. ACS Nano, 2010, 4(9): 5405-5413.

[246] Xu Y, Sheng K, Li C, et al. Self-assembled graphene hydrogel via a one-step hydrothermal process[J]. ACS Nano, 2010, 4(7): 4324-4330.

[247] Wu Z S, Sun Y, Tan Y Z, et al. Three-dimensional graphene-based macro-and mesoporous frameworks for high-performance electrochemical capacitive energy storage[J]. J. Am. Chem. Soc., 2012, 134(48): 19532-19535.

[248] Sheng K X, Sun Y Q, Li C, et al. Ultrahigh-rate supercapacitors based on eletrochemically reduced graphene oxide for ac line-filtering[J]. Sci. Rep., 2012, 2: 247.

[249] Wu Z S, Winter A, Chen L, et al. Three-dimensional nitrogen and boron co-doped graphene for high-performance all-solid-state supercapacitors[J]. Adv. Mater., 2012, 24(37): 5130-5135.

[250] Zhang X, Sui Z, Xu B, et al. Mechanically strong and highly conductive graphene aerogel and its use as electrodes for electrochemical power sources[J]. J. Mater. Chem., 2011, 21(18): 6494-6497.

[251] Zhang L, Shi G Q. Preparation of highly conductive graphene hydrogels for fabricating supercapacitors with high rate capability[J]. J. Phys. Chem. C, 2011, 115(34): 17206-17212.

[252] Chen J, Sheng K X, Luo P H, et al. Graphene hydrogels deposited in nickel foams for high-rate electrochemical capacitors[J]. Adv. Mater., 2012, 24(33): 4569-4573.

[253] Chen W F, Yan L F. In situ self-assembly of mild chemical reduction graphene

for three-dimensional architectures[J]. Nanoscale, 2011, 3(8): 3132-3137.

[254] Guo H L, Su P, Kang X, et al. Synthesis and characterization of nitrogen-doped graphene hydrogels by hydrothermal route with urea as reducing-doping agents[J]. J. Mater. Chem. A, 2013, 1(6): 2248-2255.

[255] Yuan J, Zhu J, Bi H, et al. Self-assembled hydrothermal synthesis for producing a $MnCO_3$/graphene hydrogel composite and its electrochemical properties[J]. RSC Adv., 2013, 3(13): 4400-4407.

[256] Yuan J, Zhu J, Bi H, et al. Graphene-based 3D composite hydrogel by anchoring $Co_3O_4$ nanoparticles with enhanced electrochemical properties[J]. Phys. Chem. Chem. Phys., 2013, 15(31): 12940-12945.

[257] Chen S, Duan J, Tang Y, et al. Hybrid hydrogels of porous graphene and nickel hydroxide as advanced supercapacitor materials[J]. Chem. Eur. J., 2013, 19(22): 7118-7124.

[258] Qu Q T, Shi Y, Li L L, et al. $V_2O_5 \cdot 0.6H_2O$ nanoribbons as cathode material for asymmetric supercapacitor in $K_2SO_4$ solution[J]. Electrochem. Commun., 2009, 11(6): 1325-1328.

[259] Xiong C, Aliev A E, Gnade B, et al. Fabrication of silver vanadium oxide and $V_2O_5$ nanowires for electrochromics[J]. ACS Nano, 2008, 2(2): 293-301.

[260] Perera S D, Patel B, Bonso J, et al. Vanadium oxide nanotube spherical clusters prepared on carbon fabrics for energy storage applications[J]. ACS Appl. Mater. Interfaces, 2011, 3(11): 4512-4517.

[261] Ancona M G, Kooi S E, Kruppa W, et al. Patterning of narrow au nanocluster lines using $V_2O_5$ nanowire masks and ion-beam milling[J]. Nano Letters, 2003, 3(2): 135-138.

[262] Boukhalfa S, Evanoff K, Yushin G. Atomic layer deposition of vanadium oxide on carbon nanotubes for high-power supercapacitor electrodes[J]. Energy Environ. Sci., 2012, 5: 6872-6879.

[263] Do Q H, Zeng C, Zhang C, et al. Supercritical fluid deposition of vanadium

oxide on multi-walled carbon nanotube buckypaper for supercapacitor electrode application[J]. Nanotechnology, 2011, 22(36): 365402-365409.

[264] Lampe-Önnerud C, Thomas J O, Hardgrave M, et al. The performance of single-phase $V_6O_{13}$ in the lithium/polymer electrolyte battery [J]. J. Electrochem. Soc., 1995, 142(11): 3648-3651.

[265] Shao J, Li X, Qu Q, et al. One-step hydrothermal synthesis of hexangular starfruit-like vanadium oxide for high power aqueous supercapacitors[J]. J. Power Sources, 2012, 219: 253-257.

[266] Hu L, Yu L, Zhao C, et al. Synthesis and characterization of $VO_2$/mesoporous carbon composites for hybrid capacitors[J]. J. Wuhan Univ. Technol., Mater. Sci. Ed., 2010, 25(4): 574-578.

[267] Guo C X, Li C M. A self-assembled hierarchical nanostructure comprising carbon spheres and graphene nanosheets for enhanced supercapacitor performance[J]. Energy Environ. Sci., 2011, 4(11): 4504-4507.

[268] Kovtyukhova N I, Ollivier P J, Martin B R, et al. Layer-by-layer assembly of ultrathin composite films from micron-sized graphite oxide sheets and polycations[J]. Chem. Mater., 1999, 11(3): 771-778.

[269] Shi S, Cao M, He X, et al. Surfactant-assisted hydrothermal growth of single-crystalline ultrahigh-aspect-ratio vanadium oxide nanobelts[J]. Cryst. Growth Des., 2007, 7(9): 1893-1897.

[270] Xin S, Guo Y G, Wan L J. Nanocarbon networks for advanced rechargeable lithium batteries[J]. Acc. Chem. Res., 2012, 45(10): 1759-1769.

[271] Baudrin E, Sudant G, Larcher D, et al. Preparation of nanotextured $VO_2$[B] from vanadium oxide aerogels[J]. Chem. Mater., 2006, 18(18): 4369-4374.

[272] Fang W C. Synthesis and electrochemical characterization of vanadium oxide/carbon nanotube composites for supercapacitors[J]. J. Phys. Chem. C, 2008, 112(30): 11552-11555.

[273] Pan X, Zhao Y, Ren G, et al. Highly conductive $VO_2$ treated with hydrogen

for supercapacitors[J]. Chem. Commun., 2013, 49(38): 3943-3945.

[274] Yang S, Gong Y, Liu Z, et al. Bottom-up approach toward single-crystalline $VO_2$-graphene ribbons as cathodes for ultrafast lithium storage[J]. Nano Lett., 2013, 13(4): 1596-1601.

[275] Yan J, Wang Q, Wei T, et al. Recent advances in design and fabrication of electrochemical supercapacitors with high energy densities[J]. Adv. Energy Mater., 2014, 4: 1300816.

[276] Qu Q T, Shi Y, Tian S, et al. A new cheap asymmetric aqueous supercapacitor: Activated carbon//$NaMnO_2$ [J]. J. Power Sources, 2009, 194(2): 1222-1225.

[277] Xu Y, Lin Z, Huang X, et al. Functionalized graphene hydrogel-based high-performance supercapacitors[J]. Adv. Mater., 2013, 25(40): 5779-5784.

[278] Liu Y Y, Clark M, Zhang Q F, et al. $V_2O_5$ nano-electrodes with high power and energy densities for thin film Li-ion batteries[J]. Adv. Energy Mater., 2011, 1(2): 194-202.

[279] Wei T Y, Chen C H, Chien H C, et al. A cost-effective supercapacitor material of ultrahigh specific capacitances: spinel nickel cobaltite aerogels from an epoxide-driven sol-gel process[J]. Adv. Mater., 2010, 22(3): 347-351.

[280] Soin N, Roy S S, Mitra S K, et al. Nanocrystalline ruthenium oxide dispersed Few Layered Graphene (FLG) nanoflakes as supercapacitor electrodes[J]. J. Mater. Chem., 2012, 22(30): 14944-14950.

[281] Justin P, Meher S K, Rao G R. Tuning of capacitance behavior of NiO using anionic, cationic, and nonionic surfactants by hydrothermal synthesis[J]. J. Phys. Chem. C, 2010, 114(11): 5203-5210.

[282] Luo D, Zhang G, Liu J, et al. Evaluation criteria for reduced graphene oxide [J]. J. Phys. Chem. C, 2011, 115(23): 11327-11335.

[283] Yang Q, Lu Z, Chang Z, et al. Hierarchical $Co_3O_4$ nanosheet@nanowire arrays with enhanced pseudocapacitive performance[J]. RSC Adv., 2012, 2(4):

1663 - 1668.

[284] Marconcini P, Cresti A, Triozon F, et al. Atomistic boron-doped graphene field-effect transistors: a route toward unipolar characteristics[J]. ACS Nano, 2012, 6(9): 7942 - 7947.

[285] Stoller M D, Magnuson C W, Zhu Y, et al. Interfacial capacitance of single layer graphene[J]. Energy Environ. Sci., 2011, 4(11): 4685 - 4689.

[286] Zhi M, Xiang C, Li J, et al. Nanostructured carbon-metal oxide composite electrodes for supercapacitors: a review[J]. Nanoscale, 2013, 5(1): 72 - 88.

[287] Cheng Q, Tang J, Ma J, et al. Graphene and nanostructured $MnO_2$ composite electrodes for supercapacitors[J]. Carbon, 2011, 49(9): 2917 - 2925.

[288] Cheng Q, Tang J, Ma J, et al. Graphene and carbon nanotube composite electrodes for supercapacitors with ultra-high energy density[J]. Phys. Chem. Chem. Phys., 2011, 13(39): 17615 - 17624.

[289] Goh M S, Pumera M. Multilayer graphene nanoribbons exhibit larger capacitance than their few-layer and single-layer graphene counterparts[J]. Electrochem. Commun., 2010, 12(10): 1375 - 1377.

[290] Tang W, Liu L, Tian S, et al. Aqueous supercapacitors of high energy density based on $MoO_3$ nanoplates as anode material[J]. Chem. Commun., 2011, 47(36): 10058 - 10060.

[291] Qu Q, Zhu Y, Gao X, et al. Core-shell structure of polypyrrole grown on $V_2O_5$ nanoribbon as high performance anode material for supercapacitors[J]. Adv. Energy Mater., 2012, 2(8): 950 - 955.

[292] Lu X, Wang G, Zhai T, et al. Stabilized TiN nanowire arrays for high-performance and flexible supercapacitors[J]. Nano Lett., 2012, 12(10): 5376 - 5381.

[293] Glushenkov A M, Hulicova-Jurcakova D, Llewellyn D, et al. Structure and capacitive properties of porous nanocrystalline VN prepared by temperature-programmed ammonia reduction of $V_2O_5$[J]. Chem. Mater., 2010, 22(3):

914-921.

[294] Yuan D, Zeng J, Kristian N, et al. $Bi_2O_3$ deposited on highly ordered mesoporous carbon for supercapacitors[J]. Electrochem. Commun. 2009, 11(2): 313-317.

[295] Wu M S, Lee R H, Jow J J, et al. Nanostructured iron oxide films prepared by electrochemical method for electrochemical capacitors[J]. Electrochem. Solid-State Lett., 2009, 12(1): A1-A4.

[296] Chung K W, Kim K B, Han S H, et al. Novel synthesis and electrochemical characterization of nano-sized cellular $Fe_3O_4$ thin film[J]. Electrochem. Solid-State Lett., 2005, 8(5): A259-A262.

[297] Brousse T, Bélanger D. A hybrid $Fe_3O_4$-$MnO_2$ capacitor in mild aqueous electrolyte[J]. Electrochem. Solid-State Lett., 2003, 6(11): A244-A248.

[298] Kulal P M, Dubal D P, Lokhande C D, et al. Chemical synthesis of $Fe_2O_3$ thin films for supercapacitor application[J]. J. Alloys Compd., 2011, 509(5): 2567-2571.

[299] Xie K, Li J, Lai Y, et al. Highly ordered iron oxide nanotube arrays as electrodes for electrochemical energy storage[J]. Electrochem. Commun., 2011, 13(6): 657-660.

[300] Jin W H, Cao G T, Sun J Y. Hybrid supercapacitor based on $MnO_2$ and columned FeOOH using $Li_2SO_4$ electrolyte solution[J]. J. Power Sources, 2008, 175(1): 686-691.

[301] Chen J, Huang K, Liu S. Hydrothermal preparation of octadecahedron $Fe_3O_4$ thin film for use in an electrochemical supercapacitor[J]. Electrochim. Acta, 2009, 55(1): 1-5.

[302] Wang L, Ji H, Wang S, et al. Preparation of $Fe_3O_4$ with high specific surface area and improved capacitance as a supercapacitor[J]. Nanoscale, 2013, 5(9): 3793-3799.

[303] Mu J, Chen B, Guo Z, et al. Highly dispersed $Fe_3O_4$ nanosheets on one-

dimensional carbon nanofibers: Synthesis, formation mechanism, and electrochemical performance as supercapacitor electrode materials[J]. Nanoscale, 2011, 3(12): 5034 - 5040.

[304] Guan D, Gao Z, Yang W, et al. Hydrothermal synthesis of carbon nanotube/cubic $Fe_3O_4$ nanocomposite for enhanced performance supercapacitor electrode material[J]. Materials Science and Engineering B, 2013, 178(10): 736 - 743.

[305] Liu D, Wang X, Wang X, et al. Ultrathin nanoporous $Fe_3O_4$ - carbon nanosheets with enhanced supercapacitor performance[J]. J. Mater. Chem. A, 2013, 1(6): 1952 - 1955.

[306] Wu N L, Wang S Y, Han C Y, et al. Electrochemical capacitor of magnetite in aqueous electrolytes[J]. J. Power Sources, 2003, 113(1): 173 - 178.

[307] Lee K K, Deng S, Fan H M, et al. $\alpha-Fe_2O_3$ nanotubes-reduced graphene oxide composites as synergistic electrochemical capacitor materials[J]. Nanoscale, 2012, 4(9): 2958 - 2961.

[308] Sassin M B, Mansour A N, Pettigrew K A, et al. Electroless deposition of conformal nanoscale iron oxide on carbon nanoarchitectures for electrochemical charge storage[J]. ACS Nano, 2010, 4(8): 4505 - 4514.

[309] Wang Q, Jiao L, Du H, et al. $Fe_3O_4$ nanoparticles grown on graphene as advanced electrode materials for supercapacitors[J]. J. Power Sources, 2014, 245: 101 - 106.

[310] Low Q X, Ho G W. Facile structural tuning and compositing of iron oxide-graphene anode towards enhanced supacapacitive performance[J]. Nano Energy, 2014, 5: 28 - 35.

[311] Wang H, Yi H, Chen X, et al. One-step strategy to three-dimensional graphene/$VO_2$ nanobelt composite hydrogels for high performance supercapacitors[J]. J. Mater. Chem. A, 2014, 2(4): 1165 - 1173.

[312] Liu C, Yu Z, Neff D, et al. Graphene-based supercapacitor with an ultrahigh energy density[J]. Nano Lett., 2010, 10(12): 4863 - 4868.

[313] Si Y C, Samulski E T. Exfoliated graphene separated by platinum nanoparticles [J]. Chem. Mater., 2008, 20(21): 6792-6797.

[314] Hang B T, Watanabe T, Eashira M, et al. The electrochemical properties of $Fe_2O_3$-loaded carbon electrodes for iron-air battery anodes[J]. J. Power Sources, 2005, 150: 261-271.

[315] Lee J W, Hall A S, Kim J D, et al. A facile and template-free hydrothermal synthesis of $Mn_3O_4$ nanorods on graphene sheets for supercapacitor electrodes with long cycle stability[J]. Chem. Mater., 2012, 24(6): 1158-1164.

[316] Wei W, Cui X, Chen W, et al. Manganese oxide-based materials as electrochemical supercapacitor electrodes[J]. Chem. Soc. Rev., 2011, 40(3): 1697-1721.

[317] Wang H, Casalongue H S, Liang Y, et al. $Ni(OH)_2$ nanoplates grown on graphene as advanced electrochemical pseudocapacitor materials[J]. J. Am. Chem. Soc., 2010, 132(21): 7472-7477.

[318] Guan C, Xia X, Meng N, et al. Hollow core-shell nanostructure supercapacitor electrodes: gap matters[J]. Energy Environ. Sci., 2012, 5(10): 9085-9090.

[319] Wei W, Cui X, Chen W, et al. Phase-controlled synthesis of $MnO_2$ nanocrystals by anodic electrodeposition: implications for high-rate capability electrochemical supercapacitors [J]. J. Phys. Chem. C, 2008, 112(38): 15075-15083.

[320] Yu G, Hu L, Vosgueritchian M, et al. Solution-processed graphene/$MnO_2$ nanostructured textiles for high-performance electrochemical capacitors[J]. Nano Lett., 2011, 11(7): 2905-2911.

[321] Li Y, Fu Z Y, Su B L. Hierarchically structured porous materials for energy conversion and storage[J]. Adv. Funct. Mater., 2012, 22(22): 4634-4667.

[322] Aravindan V, Chuiling W, Reddy M V, et al. Carbon coated nano-$LiTi_2(PO_4)_3$ electrodes for non-aqueous hybrid supercapacitors[J]. Phys. Chem. Chem. Phys., 2012, 14(16): 5808-5814.

[323] Wang H, Liang Y, Gong M, et al. An ultrafast nickel-iron battery from strongly coupled inorganic nanoparticle/nanocarbon hybrid materials[J]. Nat. Commun., 2012, 3: 917.

[324] Luo J, Liu J, Zeng Z, et al. Three-dimensional graphene foam supported $Fe_3O_4$ lithium battery anodes with long cycle life and high rate capability[J]. Nano Lett., 2013, 13(12): 6136-6143.

[325] Linden D, Reddy T B. Handbook of Batteries[M]. NewYork: McGraw-Hill, 1995.

[326] Wang H, Yi H, Chen X, et al. Asymmetric supercapacitors based on nano-architectured nickel oxide/graphene foam and hierarchical porous nitrogen-doped carbon nanotubes with ultrahigh-rate performance[J]. J. Mater. Chem. A, 2014, 2(9): 3223-3230.

[327] Cong H P, Ren X C, Wang P, et al. Macroscopic multifunctional graphene-based hydrogels and aerogels by a metal ion induced self-assembly process[J]. ACS Nano, 2012, 6(3): 2693-2703.

[328] Zhao Y, Hu C, Hu Y, et al. A versatile, ultralight, nitrogen-doped graphene framework[J]. Angew. Chem., 2012, 124(45): 11533-11537.

[329] Feng D, Lv Y, Wu Z, et al. Free-standing mesoporous carbon thin films with highly ordered pore architectures for nanodevices[J]. J. Am. Chem. Soc., 2011, 133(38): 15148-15156.

[330] Liu X, Jung H G, Kim S O, et al. Silicon/copper dome-patterned electrodes for high-performance hybrid supercapacitors[J]. Sci. Rep., 2013, 3: 3183.

[331] Plitz I, DuPasquier A, Badway F, et al. The design of alternative nonaqueous high power chemistries[J]. Appl. Phys. A, 2006, 82(4): 615-626.

[332] Pasquier A D, Plitz I, Menocal S, et al. A comparative study of Li-ion battery, supercapacitor and nonaqueous asymmetric hybrid devices for automotive applications[J]. J. Power Sources, 2003, 115(1): 171-178.

[333] Stewart S, Albertus P, Srinivasan V, et al. Optimizing the performance of

lithium titanate spinel paired with activated carbon or iron phosphate[J]. J. Electrochem. Soc., 2008, 155(3): A253 - A261.

[334] Aravindan V, Chuiling W, Madhavi S. High power lithium-ion hybrid electrochemical capacitors using spinel LiCrTiO$_4$ as insertion electrode[J]. J. Mater. Chem., 2012, 22(31): 16026 - 16031.

[335] Wang Q, Wen Z, Li J. A hybrid supercapacitor fabricated with a carbon nanotube cathode and a TiO$_2$ - B nanowire anode[J]. Adv. Funct. Mater., 2006, 16(16): 2141 - 2146.

[336] Ni J, Yang L, Wang H, et al. A high-performance hybrid supercapacitor with Li$_4$Ti$_5$O$_{12}$ - C nano-composite prepared by in situ and ex situ carbon modification [J]. J. Solid State Electrochem., 2012, 16(8): 2791 - 2796.

[337] Stoller M D, Murali S, Quarles N, et al. Activated graphene as a cathode material for Li-ion hybrid supercapacitors[J]. Phys. Chem. Chem. Phys., 2012, 14(10): 3388 - 3391.

[338] Aravindan V, Mhamane D, Ling W C, et al. Nonaqueous lithium-ion capacitors with high energy densities using trigol-reduced graphene oxide nanosheets as cathode-active material[J]. ChemSusChem, 2013, 6(12): 2240 - 2244.

[339] Jain A, Aravindan V, Jayaraman S, et al. Activated carbons derived from coconut shells as high energy density cathode material for Li-ion capacitors[J]. Sci. Rep., 2013, 3: 3002.

[340] Li Y G, Tan B, Wu Y Y. Mesoporous Co$_3$O$_4$ nanowire arrays for lithium ion batteries with high capacity and rate capability[J]. Nano Lett., 2008, 8(1): 265 - 270.

[341] Shen L, Uchaker E, Zhang X, et al. Hydrogenated Li$_4$Ti$_5$O$_{12}$ nanowire arrays for high rate lithium ion batteries [J]. Adv. Mater., 2012, 24(48): 6502 - 6506.

[342] Luo Y, Luo J, Jiang J, et al. Seed-assisted synthesis of highly ordered TiO$_2$@ α - Fe$_2$O$_3$ core/shell arrays on carbon textiles for lithium-ion battery applications

[J]. Energy Environ. Sci., 2012, 5(4): 6559-6566.

[343] Hulicova-Jurcakova D, Puziy A M, Poddubnaya O I, et al. Highly stable performance of supercapacitors from phosphorus-enriched carbons[J]. J. Am. Chem. Soc., 2009, 131(14): 5026-5027.

[344] Cao Y, Xiao L, Sushko M L, et al. Sodium ion insertion in hollow carbon nanowires for battery applications[J]. Nano Lett., 2012, 12(7): 3783-3787.

[345] Chen S, Zhu J, Wu X, et al. Graphene oxide-$MnO_2$ nanocomposites for supercapacitors[J]. ACS Nano, 2010, 4(5): 2822-2830.

[346] Yuan Q, Hu H, Gao J, et al. Upright standing graphene formation on substrates[J]. J. Am. Chem. Soc., 2011, 133(40): 16072-16079.

[347] Shen Y D, Li Y W, Li W M, et al. Growth of $Bi_2O_3$ ultrathin films by atomic layer deposition[J]. J. Phys. Chem. C, 2012, 116(5): 3449-3456.

[348] Allen M J, Tung V C, Kaner R B. Honeycomb carbon: a review of graphene [J]. Chem. Rev., 2010, 110(1): 132-145.

[349] Zhao X, Tian H, Zhu M, et al. Carbon nanosheets as the electrode material in supercapacitors[J]. J. Power Sources, 2009, 194(2): 1208-1212.

[350] Aravindan V, Shubha N, Chui Ling W, et al. Constructing high energy density non-aqueous Li-ion capacitors using monoclinic $TiO_2$-B nanorods as insertion host[J]. J. Mater. Chem. A, 2013, 1(20): 6145-6151.

[351] Wang C, Zhang X, Zhang Y, et al. Hydrothermal growth of layered titanate nanosheet arrays on titanium foil and their topotactic transformation to heterostructured $TiO_2$ photocatalysts[J]. J. Phys. Chem. C, 2011, 115(45): 22276-22285.

[352] Wang D, Choi D, Li J, et al. Self-assembled $TiO_2$-graphene hybrid nanostructures for enhanced Li-ion insertion[J]. ACS Nano, 2009, 3(4): 907-914.

[353] Chen Z, Augustyn V, Wen J, et al. High-performance supercapacitors based on intertwined $CNT/V_2O_5$ nanowire nanocomposites[J]. Adv. Mater., 2011,

23(6): 791-795.

[354] Park J H, Park O O. Hybrid electrochemical capacitors based on polyaniline and activated carbon electrodes[J]. J. Power Sources, 2002, 111(1): 185-190.

[355] Bi H, Huang F, Liang J, et al. Large-scale preparation of highly conductive three dimensional graphene and its applications in CdTe solar cells[J]. J. Mater. Chem., 2011, 21(43): 17366-17370.

[356] Luo J, Xia X, Luo Y, et al. Rationally designed hierarchical $TiO_2$@$Fe_2O_3$ hollow nanostructures for improved lithium ion storage[J]. Adv. Energy Mater., 2013, 3(6): 737-743.

[357] Aravindan V, Cheah Y L, Mak W F, et al. Fabrication of high energy-density hybrid supercapacitors using electrospun $V_2O_5$ nanofibers with a self-supported carbon nanotube network[J]. ChemPlusChem, 2012, 77(7): 570-575.

[358] Wang G, Liu Z Y, Wu J N, et al. Preparation and electrochemical capacitance behavior of $TiO_2$-B nanotubes for hybrid supercapacitor[J]. Mater. Lett., 2012, 71: 120-122.

[359] Karthikeyan K, Amaresh S, Lee S N, et al. Fluorine-doped $Fe_2O_3$ as high energy density electroactive material for hybrid supercapacitor applications[J]. Chem. Asian J., 2014, 9(3): 852-857.

[360] Aravindan V, Reddy M V, Madhavi S, et al. Hybrid supercapacitor with nano-$TiP_2O_7$ as intercalation electrode[J]. J. Power Sources, 2011, 196(20): 8850-8854.

[361] Kim H, Cho M Y, Kim M H, et al. A novel high-energy hybrid supercapacitor with an anatase $TiO_2$-reduced graphene oxide anode and an activated carbon cathode[J]. Adv. Energy Mater., 2013, 3(11): 1500-1506.

# 后 记

  首先我要衷心地感谢我的导师王雪峰教授三年来在学业上给予我的悉心指导。在博士期间,王老师给我营造了轻松自由的科研环境,也同时给我提供了国外交流的学术平台。导师严谨的治学态度、渊博的专业知识、忘我的工作精神、国际化的学术视野和对学生的关心爱护给我留下了积极而深远的影响,让我受益匪浅。正是导师多年的教育和培养,将我引入了科学研究的前沿领域,使我开阔了眼界、拓宽了知识面。在这三年中,我不仅学习了系统研究的科学方法,还学会了许多做人的道理,这些都将成为我今后人生中永久的财富。在此,我谨向王老师表示衷心的感谢。

  感谢新加坡南洋理工大学的范红金(FAN Hongjin)教授给我提供了"国外短期访学"的机会,让我开阔了视野、增长了见识。同时,感谢复旦大学化学系的傅正文教授和刘伟明博士给予我锂电池方面的指导。感谢许兵老师在博士期间给予的关心和帮助。

  同时,实验中心的各位老师给予了热情的关心和指导,在测样方面提供了许多方便,在此一并向他们表示诚挚的谢意。

  感谢实验室的王强、刘星、易欢、荆语婷、赵杰博士、王雅兰、仝丽洁、李广君、陈晓、刘小兵、潘震云、瞿淼、张晨曦、宁超、刘昊、俞文杰、彭天权等师兄师弟师妹们在学习和生活中给予的支持和帮助。

  同时,我将最真诚的谢意献给我的父母和家人,他们的支持和期待给了我完成学业无尽的力量。

王欢文